Geopedia

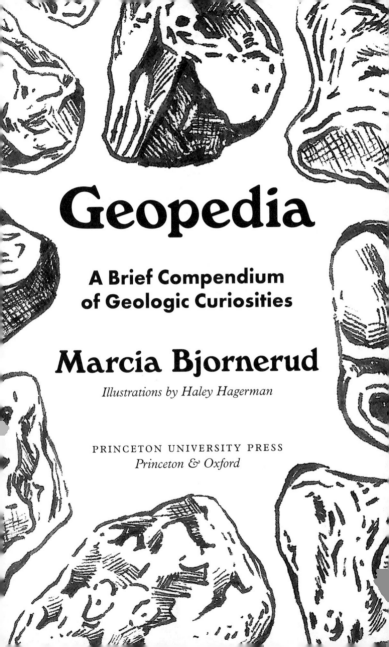

Geopedia

A Brief Compendium of Geologic Curiosities

Marcia Bjornerud

Illustrations by Haley Hagerman

PRINCETON UNIVERSITY PRESS
Princeton & Oxford

Published by Princeton University Press
41 William Street, Princeton, New Jersey 08540
6 Oxford Street, Woodstock, Oxfordshire OX20 1TR

press.princeton.edu

All Rights Reserved
ISBN 9780691212579
ISBN (e-book) 9780691232720

British Library Cataloging-in-Publication Data is available

Editorial: Robert Kirk and Abigail Johnson
Production Editorial: Mark Bellis
Text and Cover Design: Chris Ferrante
Production: Steve Sears
Publicity: Sara Henning-Stout and Caitlyn Robson
Copyeditor: Cathryn Slovensky

Cover, endpaper, and text illustrations by Haley Hagerman

This book has been composed in Plantin, Futura, and Windsor

Printed on acid-free paper. ∞

Printed in China

10 9 8 7 6 5 4 3 2 1

For F, G, J, K, O, and P
with love from M

Contents

Preface

Hello, Earthling.

You might not be in the habit of thinking of yourself as such, but *Earthling* is your most fundamental identity. You have deep evolutionary roots in this planet. You are literally made of Earth—of water that has cycled for eons through clouds and rivers and oceans, and of minerals in the soil, derived from rock that was itself forged from the planet's interior. In modern times, technology and urban infrastructure create the illusion that we have gained autonomy from the natural world, but our ancestors were deeply aware of our underlying earthiness. In Hebrew, "Adam" means "earth" or "clay," and the word "human" shares an ancient Indo-European root with *humus* or soil—a profound acknowledgment of our essential nature.

If we were formed from the stony ground, stone also made us human. It was the medium of our first and longest technological age. In fact, we are still very much in the stone age, utterly dependent on rocks as a source of groundwater, building materials, fossil fuels, metals, elements for high-tech devices—and every other commodity that can't be cultivated, raised, or hunted.

Yet most of us give little thought to this geologic infrastructure or the workings of the planet—mainly because few have had the opportunity to get to know the components of the earth by name. Even at schools with otherwise strong science offerings in physics, chemistry, and biology, rigorous courses in the geosciences are rare, and for schools with fewer resources, geology is seen as nonessential. As a consequence, we are a geologically illiterate society, a condition that has not only led to unwitting environmental degradation but has also cut us off from a sense of our shared heritage as offspring of the earth.

All of this is an unfortunate happenstance of intellectual history; among the sciences, geology is a bit of a late bloomer. In the early 19th century, when physicists were discovering principles that promised new levels of human mastery over matter, geologists were mostly accumulators of odd stones, owners of curiosity cabinets, curators of museum displays. And while Victorian-era geologists did a thorough job of describing the anatomy of the earth (its fossil-bearing strata, rocks and minerals, surface features), the complex *physiology* of the planet—plate tectonics, the climate system, global biogeochemical cycles—remained largely unknown until the mid- to late twentieth century. By that time, the fusty reputation of geology as a science focused on collection and classification of inert artifacts from the murky past was well established in the public mind.

The outdated public perception of the discipline frustrates present-day earth scientists, because the last few decades have, in fact, been a golden time for geology—or geoscience, to use a term that embraces not only the study of rocks but also of the atmosphere, oceans, ice caps, magnetic field, and other moving parts in the earth system, and not just the planet's past but also its present and future. Modern geoscience combines field observations, which have been the foundation of the discipline since the 19th century, with high-precision geochemical analyses, satellite observations, geophysical monitoring, numerical modeling, and other techniques that make it possible to understand Earth processes over timescales ranging from the seconds in which an earthquake occurs to the entire 4.5-billion-year history of the planet. To geoscientists, rocks are not nouns but verbs—far more than inert curios, they are evidence of Earth's ebullient creativity, its capacity for ceaseless reincarnation of primordial matter into new forms. Rocks are transcripts of eons of conversation between the solid earth and water, air and life. They are both fascinating archives of the past and our best windows into the future. And the geologic lexicon reflects not only the prodigious diversity of rocks and geologic phenomena but also the rich history of human experiences with them over the last 10 millennia.

I understand that for outsiders, geologic terminology can at first be opaque and off-putting. Chemistry, at least, has consistent rules about naming compounds, and biology uses Linnaean taxonomy to impose order on the unruly multitudes of organisms. The technical glossary of geology, in contrast, is a gallimaufry of terms from mythology, neologisms concocted from Greek and Latin roots, embarrassing anachronisms, and utilitarian recent coinages. It also includes words imported from scores of world languages, ranging from Arabic (*erg*) to Inuktitut (*nunatak*), Slovenian (*karst*) to Javanese (*lahar*), based on the premise that people who have direct experience of a geologic phenomenon are in the best position to describe it. Perhaps geoscientists can be forgiven for being such magpies for words; the sheer profusion of things invented by this creative planet demands an immense vocabulary to match.

This short book is certainly not intended as a systematic introduction to the geosciences, nor a comprehensive glossary of the field. The American Geosciences Institute publishes such a volume, and it runs to more than 39,000 entries, including more than 5,000 mineral names alone. Instead, *Geopedia* is an admittedly idiosyncratic compendium of words and terms chosen because they are portals into larger geologic stories—of remarkable places, strange incidents, dramatic plot twists in the planet's history, misconceptions about geologic phenomena, colorful characters who contributed to the geosciences, and the remarkable biographies of selected rocks, minerals, and landforms that every Earthling should know on a first-name basis. Sadly, too many Earthlings live their lives on this planet like bad tourists—taking its many amenities for granted without a thought for how these came to be, and never learning even the rudiments of the history, language, and culture of this wonderfully strange rock we call home.

Whether you read this book alphabetically, from "Acasta Gneiss" to "Zircon," or meander like a river through it by following the cross-references at the end of each entry (see *Thalweg*) or the itineraries suggested in the appendixes, my hope is that in the process you will gain an impressionistic sense of how Earth works, how it has coevolved with Life over billions of years, and how our understanding of it has deepened over time.

Welcome to the curiosity cabinet.

Acasta Gneiss [ah-CAST-ah nice]
The Old World

In a remote, roadless part of Canada's Northwest Territories, just east of Great Bear Lake, an ice-scoured stretch of somber gray and white-striped rock lies open to the subarctic sky. Across the vast expanse of the ancient Canadian Shield, there are many other outcrops like it, but to geologists, these rocks, known as the ***Acasta Gneiss Complex***, are celebrities: the oldest yet found on Earth, clocking in at the astonishing age of 4.03 billion years. Their geographic inaccessibility seems fitting, mirroring their geologic remoteness in time. It shouldn't be easy to visit these Old Ones.

Even for geologists who shuttle frequently back and forth in "Deep Time," grasping what four billion years actually means can be difficult. One way to make such an immense span seem more real is to think like the rocks and reconceive what to us is the geologic past as the one-time geologic future. This is akin to seeing a photograph of your great-grandmother when she was a child—when the course of her life, and all the events in your grand-mother's and mother's lives that led to your existence, were yet unimagined—and trying to understand how different the world, and its future, must have seemed then.

The Acasta gneiss is something like our hundred-millionth great-grandmother, who yet still dwells among us. Old Acasta remembers Earth as it was before it had the attributes that define Earth today. She was here long, long before the dinosaurs, in fact, before there were plants and animals on land, or oxygen in the at-mosphere, or possibly even microbial life—and probably before Earth had settled into the habit of plate tectonics. And none of these things—though now literally set in stone—were then preordained. Given the happen-stances of planetary and biological evolution, Earth's story could have unfolded quite differently.

The Acasta rocks are the same age as the giant impact basins on the moon. These basins—Galileo's "maria" or "seas"—formed during the Late Heavy Bombardment, a fusillade of large meteors in the inner solar system be-tween about 4.1 and 3.8 billion years ago. The rocks at Acasta survived not only this salvo but multiple episodes of deformation and recrystallization, which transformed them from their original state as granitic rocks into the zebra-striped metamorphic rock called "gneiss" (which

is as nice as it sounds). They've also endured eons of
erosion, burial, and reexhumation as plates collided, the
crust warped, seas rose and fell, rivers raged, and glaciers
waxed and waned.

Although the Acasta gneiss is immensely old, these
rocks are about 530 million (0.53 billion) years younger
than the 4.56-billion-year age of the earth itself. This is
a nontrivial expanse of time; it's about the same interval
that separates us from the dawn of animal life in the
middle Cambrian period. Any Earth rocks that formed
before the Acasta gneiss were apparently so thoroughly
altered, via melting, meteorite impacts, and some sort
of pre-plate-tectonic remixing, that we find no vestige
of them—except for a handful of tiny crystals of the durable mineral zircon, preserved in an ancient sandstone
in Western Australia.

One may then logically wonder, if no rocks survive
from the formation of the earth, how has the age of the
planet been determined? An astute question! The age
of the earth, paradoxically, comes from extraterrestrial
objects—meteorites—that formed at the same time as
Earth and the other members of the solar system but
have remained unchanged in the subsequent 4.5 billion
years while Earth has continuously resurfaced and re-
modeled itself.

As Earth's oldest surviving rock complex, the Acasta
gneiss marks the end of the Hadean eon—the first interval of the geologic timescale, defined as the time
period in Earth's past for which there is no native rock
record. But starting from the time of Great-Grandma
Acasta, Earth has kept a rich, if sometimes cryptic,
diary of its activities. Geologists are essentially just the

translators of that sprawling journal, and *Geopedia* is a collection of a few quirky quotations from its pages.

See also Anthropocene; Chondrite; Cryogenian; Zircon; Simplified Geologic Timescale (Appendix 1).

Allochthon [ah-LOCK-thon]
Rocks that roam

Literally, "foreign ground," an ***allochthon*** is a mass of rock that has been displaced laterally from its original location along a subhorizontal fault by tectonic forces. In some cases, such slabs have been shoved tens of miles off their foundations. The term has connections to Greek mythology: the chthonic deities, including Hades, Persephone, and Charon, the ferryman on the River Styx, lived underground.

Before plate-tectonic theory emerged in the 1960s, continents were believed to be rooted in place, and crustal deformation like that seen in mountain belts was thought to be driven solely by the vertical force of gravity. This made it hard to explain observations by some astute late 19th-century geologists that strata in the Alps, Canadian Rockies, and Scottish Highlands had been shoved laterally far from where they were formed—in most cases *up* the slope of gently inclined fault surfaces. This conundrum became known as the "overthrust paradox" and inspired many creative, and mostly incorrect, hypotheses until geologists finally accepted the idea that continents dance about the globe over time, sometimes colliding and causing rocks to relocate. Allochthons remind us that even sedimentary rocks are by no means sedentary.

See also Geosyncline; Klippe.

Amethyst

Purple haze

Like place-names on a map, the names of minerals are windows into earlier cultures and ways of seeing the world. ***Amethyst***, the purple semiprecious stone that is as popular among New Age crystal worshippers as it was in classical times, serves as a colorful example. It retains the name given to it by the ancient Greeks: *amethustos* or "not drunken," which was based on their belief that it allowed the wearer to drink wine without becoming intoxicated (it seems that a few simple empirical tests might have disproved this hypothesis).

Amethyst also illustrates the scientific challenges in giving names to the prodigious variety of mineral species on Earth. The situation is akin to the parallel systems of common and Linnaean names for organisms, but in the case of minerals, there are not only folk or traditional names but also commercial gemological names and technical scientific names—and in many cases what

is recognized as a distinct "species" under one of these systems is lumped into a group by another.

The technical definition of a mineral, as decreed by the International Mineralogical Association (IMA)—the governing body that maintains law and order in the realm of minerals so we can all sleep at night—is "a naturally occurring inorganic substance with a particular chemical composition and defined crystal structure."

The IMA definition seems simple enough, but it masks some complicating subtleties. First, although minerals must be inorganic (coal, as decomposed plant matter, doesn't qualify), a large number are biogenic—that is, direct or indirect products of living organisms. Most of the calcite in limestones, for instance, was precipitated by tiny marine organisms. (Parenthetically but crucially, this process locks volcanic carbon dioxide away in solid form and is to be thanked for keeping Earth from becoming a greenhouse planet; see *Stylolite*.) More generally, many oxide minerals like hematite (Fe_2O_3) would not exist without the atmospheric oxygen produced by photosynthesizing phytoplankton and plants.

Earth's inventory of mineral species—more than 5,700 by the IMA's reckoning—has actually grown and changed over time. There are only a few hundred minerals on the arid, airless, barren moon, where volcanism ended billions of years ago and the only ongoing processes are occasional meteorite strikes and relentless irradiation by the solar wind. Earth's mineral diversity, in contrast, reflects the sheer profusion of agents—geological, hydrological, atmospheric, and biological—that incessantly dismantle and reassemble the raw materials of the planet into new forms.

Another complication in the technical definition of minerals is the matter of "particular chemical composition." All minerals have "official" chemical formulas with well-defined ratios of their constituent elements. Calcite, for example, is calcium carbonate, or $CaCO_3$. But in virtually all minerals, there are elemental impurities and substitutions in the crystal lattice. Most naturally occurring calcite, for example, has some magnesium and iron atoms sitting in the "seats" where calcium should be. Even very small amounts of such impurities may radically alter the color of a mineral. This phenomenon of "ionic substitution" is ubiquitous in the mineral kingdom and means that the number of minerals is effectively infinite. So to avoid utter taxonomical anarchy, the IMA has defined clear compositional ranges for minerals.

By this reckoning, amethyst does not qualify as a distinct mineral and is not on the official IMA list of approved names. It is simply a subvariety of quartz (SiO_2), in which trace quantities of iron and other metals that stole into silicon sites at the time of crystallization reveal their presence rather flamboyantly through the purple hue.

Similarly, the IMA does not recognize aquamarine or emerald (both subvarieties of beryl) nor ruby and sapphire (both types of corundum, a very tough mineral whose less glamorous varieties are used in sandpaper). These colorful jewels are very real, however, to those in the precious stone business and have legal definitions enforced by the IMA's commercial counterparts, the Gemological Institute of America (GIA) and the International Gemological Institute (IGI).

And who is to say which system is correct? "The Beryl Isle" just doesn't sound as verdant, "corundum slippers" don't seem as magical, and mere quartz pales next to wine-tinged amethyst.

See also Kimberlite; Pedogenesis.

Amygdule [ah-MIG-dyol]
On the bubble

An **amygdule**, from a Greek root meaning "almond," is a mineral-filled void in a porous volcanic rock, most typically basalt (the black lava of Hawai'i and Iceland). They're only vaguely almond-like in size and shape, but what they do have in common with almonds is that they

require plenty of groundwater to grow. Cultivating an amygdule, however, requires a fiery start.

Volcanic lavas are complex three-phase mixtures of liquid (magma), solid (crystals), and gases (mainly water vapor, carbon dioxide, and sulfur dioxide). As erupted lava spreads over the land surface, the gases within it escape upward, causing the tops of solidified lava flows to have a foamy texture, rather like a petrified version of the head on a glass of beer. The individual holes in this rocky foam are called "vesicles," and when groundwater—which invariably carries elements in solution—later moves through these vesicles, it may gradually fill them with a variety of new minerals, forming multicolored amygdules.

The famous agates of the North Shore of Lake Superior, with their concentric stripes of red, brown, gold, and white are amygdules formed in this way. Their polychromatic banding is a visible archive of ancient groundwater chemistry, more like the rings of an old almond tree than the fruit of a single season.

See also Brimstone; Porphyry; Nuée Ardente.

Anthropocene [AN-thrup-oh-seen]
It's about time
In contrast to measures of time like seconds, days, and years, whose durations are uniform and definitions precise, units of geologic time—ages, epochs, periods, eras, and eons—are variable in length. Indeed, the main divisions of the geologic timescale were established by paleontologists in the mid-1800s, decades before the discovery of radioactive decay made it possible to determine the absolute ages of rocks.

But these time boundaries are anything but arbitrary. They mark the beginnings and endings of great chapters in Earth's story—major inflection points in the planet's evolution—such as the rise of oxygen in the atmosphere, the dawn of animal life, and mass extinctions of varying severity. And in this century, many geoscientists think that the planetary-scale effects of human activities represent another such inflection point.

Although it is not yet an official division of the geologic timescale, the idea of the ***Anthropocene***—a new epoch in which humans have become agents of planetary change—has already gained traction both in the scientific community and the public sphere. The arguments for the Anthropocene are compelling; among them are the sobering facts that we humans now emit fifty times more carbon dioxide each year than do all of Earth's volcanoes, and we move an order of magnitude more sediment than all the rivers in the world. We have altered at least 70 percent of Earth's ice-free land surface and created vast oceanic dead zones that threaten the entire marine biosphere.

There are also legitimate critiques of the Anthropocene, however. For one thing, not all members of our species are equally culpable for the environmental damage inflicted on the planet; citizens of affluent, industrialized societies are responsible for a grossly disproportionate share of Anthropocene degradation, so why should all of humanity now share the blame?

And there is the academic matter of how exactly to define the start of the Anthropocene—the moment when we lost our geologic innocence and stepped across the threshold into a brave new world with different planetary rules. Other boundaries in geologic time have physical representations in the rock record—most infamously, the worldwide iridium layer left by the meteorite that ended the reign of the dinosaurs, marking the close of the Mesozoic era and the dawn of the Cenozoic. To what stratum do we point for the onset of Anthropocene time? Perhaps the shameful level in the Greenland and Antarctic ice sheets recording our hubristic splitting of atoms—but that ice may not long persist, owing to our insouciant alteration of the climate, the primary hallmark of the Anthropocene.

Even without a formal scientific definition, the Anthropocene is at least a reminder to Earthlings who otherwise give little thought to the planet's long history that we, too, live in geologic time. The calibration of Deep Time is one of the greatest, but least celebrated, intellectual achievements of humankind—a massive, intergenerational, global scientific project. Each time I teach History of Earth and Life, a course for geoscience majors that chronicles the full 4.5-billion-year sweep of geologic time, I marvel again that we have been able to reconstruct in such detail so much of the planet's byzantine story. (When beloved colleagues of mine complain that they can't possibly cover Renaissance art or 17th-century Russian history in one academic term, it's hard for me to feel sympathy.)

As a human creation that itself evolved over time, the geologic timescale is a rich and idiosyncratic cultural artifact. Many of its divisions are named for places where representative rocks of various ages were first systematically described. The Cambrian period, for example, is a nod to the slates of Wales ("Cambria" being the

Roman corruption of the Welsh "Cymru"); "Devonian" was originally a specific reference to the marmalade-colored sandstones of the county otherwise famous for its lavish cream teas.

Our own Quaternary period is a relic of an 18th-century timescale in which rocks were identified in a simple ordinal scheme as Primary (or Primordial), Secondary, Tertiary, and Quaternary. The International Stratigraphic Commission, the august body with jurisdiction over Deep Time, finally abolished the Tertiary in 2013 (replacing it with two shorter periods, the Paleogene and Neogene), but for now Quaternary remains as a charming anachronism. The Quaternary is further divided into Pleistocene ("mostly recent" time, the Ice Age) and Holocene history ("wholly recent"—the last 10,000 years, or all of human history). And now the Anthropocene may, ignominiously, join this list.

Perhaps the real question is not when the Anthropocene began but when it will end. A useful cultural unit of time is the *saeculum*—broadly, the time between a major event like a war or an epidemic and the death of

the last person who had firsthand memories of it. The Anthropocene has already lasted a saeculum; no one alive today remembers the world before humans profoundly altered the planet. And there are two ways for the Anthropocene to draw to a close: either we learn to be better Earthlings—blending into the background and no longer distorting biogeochemical cycles—or we go extinct. In the latter case, the Anthropocene will last a geologic saeculum: the time it will take for the earth to forget we ever existed.

See also Unconformity; Uniformitarianism; Simplified Geologic Timescale (Appendix 1).

Areology [air-ee-AH-loh-gee]
Wars of the Worlds

Areology is the rather ungainly term for the study of the "geology" of Mars, a modern homage to the ancient Greek warrior god Ares (counterpart to the Romans' Mars). Since *geo*logy literally means "the science of Earth," the term isn't technically correct when applied to other planets or moons. The study of Earth's moon is sometimes called "selenology" for the Greek moon goddess Selene, a term that peaked in popularity in the late 1960s in the lead-up to the Apollo moon landings. ("Lunology" probably sounded a little too close to lunacy for NASA's public relations team.) No parallel term has yet gained currency for Venus or Mercury (the other rocky or "terrestrial" planets), nor for the rocky moons of the outer gas giant planets.

Although Earth and Mars (and the other planets) share an origin 4.5 billion years ago in the formation of the solar system, the earth-based geological timescale is not a good framework for describing the chapters in Mars's past. It doesn't make sense to use geocentric time divisions like the Mesozoic—"Middle Life," the era of the dinosaurs—when speaking of Mars. For one thing, much of the "action" recorded in the landscapes and rocks on Mars happened in the Hadean eon (4.5–4.0 billion years ago), a time for which there is no surviving rock on Earth and which therefore has no subdivisions on the geologic timescale. By the end of Earth's Archean eon (4.0–2.5 billion years ago), the infancy and childhood of our planet, things were already slowing down on Mars. Its magnetic field shut down and its fading volcanoes could no longer keep up with losses of atmospheric gases to space. At about the time that Earth's plate-tectonic system was becoming established, Mars was already beginning to slide into a cold slumber.

The main intervals in the Martian timescale are: Pre-Noachian (4.5–4.1 billion years ago), Noachian (4.1–3.7 billion), Hesperian (3.7 to 2.9 billion), and Amazonian (2.9 billion to present). If you're thinking, "Noachian as in Noah's ark?," you're right to wonder. Naming a geologic time period for Noah would have been a scientific taboo for Earth, given geologists' strenuous efforts to fend off creationists and biblical literalists over the past two centuries. The Noachian period is technically named for an ancient region in Mars's southern hemisphere, Noachis Terra, which perhaps reminded an early astronomic observer of Mount Ararat. But there is in fact evidence of cataclysmic flooding on early Mars—perhaps *jökulhlaups*, related to sudden melting of glacial ice or permafrost. Since there is no mention of Martian floods in the Bible, an allusion to an ancient religious myth is perhaps acceptable for a planet named for a classical god.

See also Acasta Gneiss; Jökulhlaup; Uniformitarianism; Simplified Geologic Timescale (Appendix 1).

Benioff-Wadati Zone

Off the deep end

It's a bit embarrassing for geologists to acknowledge that the laws of thermodynamics were worked out in the 19th century and the structure of the atom in the early 20th, while plate tectonics—the way the solid earth works—was not known until 1965! Even when I was a college student in the early 1980s, there were still many faculty members in geoscience departments around the world who themselves had been educated under a pre-plate-tectonic paradigm with now discarded theories for how mountains grow (see *Geosyncline*) and no unifying explanation for the global distribution of earthquakes and volcanoes.

Geologists are partly to blame for refusing to take seriously the compelling evidence for continental drift published by Alfred Wegener around 1915. Wegener was a persona non grata in the minds of English-speaking geologists, as he was a German national in the time of the Great War and an interloping outsider to the field (a meteorologist). But limited geophysical information also contributed to the delayed understanding of the workings of the solid earth. Without detailed bathymetric and magnetic maps of the deep ocean floor, the fundamental process of seafloor spreading could not have been documented. And without a dense global database of earthquake locations, the boundaries of the tectonic plates were not obvious.

In particular, earthquake data were essential to the discovery of subduction—the signature process of Earth's tectonic system, by which a slab of old ocean crust, much colder and denser than it was when it

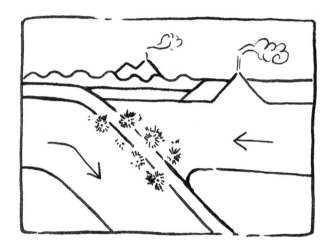

formed at a seafloor spreading ridge, is recycled back into the mantle—the fluid but solid middle layer of Earth, 1,800 miles thick, that makes up 84 percent of the planet's volume.

However, ocean-floor slabs do not go gently into the dark interior of the earth. Frictional resistance between the upper edge of the sinking plate and the lower surface of the overriding crust leads to the largest and most devastating earthquakes the planet can generate—magnitude 9 "megathrust" events. Because they invariably involve ocean crust, these monster quakes may also unleash enormous tsunamis, like those that tragically struck Sumatra in 2004 and Central Japan in 2011. Such events are (thankfully) rare enough that there were too few in the early twentieth century from which to draw generalized conclusions.

But early glimpses of the process that would be called "subduction" were detected by two pioneering seismologists working independently on opposite sides of the Pacific: Victor Hugo Benioff (1899–1968) of Cal Tech and Kiyoo Wadati (1902–95) of the Japan Meteorological Agency. Benioff and Wadati were both exceptionally skilled in deciphering information encrypted in seismograms, and they each recognized evidence for anomalously deep earthquakes at several locations around the world.

Most earthquakes occur at depths above about 10 miles, where rocks are strong and brittle, capable of failing suddenly by faulting. But Benioff and Wadati found that some earthquakes occurred in planar arrays far below this level, and in rare cases as deep as 400 miles, well into the Earth's upper mantle, where temperatures are too high for fracturing and frictional slip. Both seismologists postulated that there must be sheets of anomalously cold and/or strong rocks at these locations, but it was not until the late 1960s that their work was integrated with other geophysical observations to become modern plate theory. In honor of their prescient work, subduction zones are sometimes called ***Benioff-Wadati*** (or Wadati-Benioff) zones.

Benioff died just before the implications of his work were fully appreciated; Wadati lived long enough to enjoy accolades for his revolutionary work. Both are inspirations to those perceptive enough to discern the shape of the truth long before it comes into focus for everyone else.

See also Chondrite; Deborah Number; Eclogite; Geosyncline; Mylonite; Ophiolite.

Bioturbation

The worm churns

A portmanteau term spliced together from the words "biological perturbation," ***bioturbation*** refers to the churning of sediments on the seafloor by organisms like worms, snails, and small arthropods. An entire subdiscipline within paleontology, ichnology—from the Greek for "track" or "trace"—is dedicated to the study of burrows, trails, and other evidence of scavenging, scurrying, and excavating by ancient creatures. (Ichnology is not to be confused with "ichthyology," the study of fish, living and fossilized—although it is possible to imagine cases of bottom-dwelling, tail-dragging fish where the two specialties converge.)

Even if there were no fossils recording the emergence of animals in the Cambrian explosion about 525 million years ago, it would be clear from the physical character of marine sedimentary rocks that there were some new arrivals then on the geologic scene. In rocks deposited prior to that time, shallow water sediments commonly preserve fine, millimeter-scale laminations—in some cases, high-fidelity records of daily tidal cycles. In younger marine rocks, such layers are no longer neat and pristine but instead plowed up and ransacked, like a house in disarray following a wild party.

The diggers and grubbers that first appeared in the Cambrian also brought an end to the long reign of Earth's first complex ecosystems—the stromatolites, teeming mats of microbes that had thrived for eons in shallow coastal waters since early Archean time. After dominating the biosphere for three billion years, stromatolites went into sharp decline in the Cambrian, consigned ever after to be hors d'oeuvres for ravening invertebrates who invariably show up for the bioturbation bacchanal.

See also Cryogenian; Ediacara; Taphonomy; Simplified Geologic Timescale (Appendix 1).

Boudin [boo-DAN]
Secrets of sausage making

Field geologists are always hungry, so it's not surprising that some rock features, though quite inedible, are named for foods. One example is ***boudin***, from the French (and Cajun) word for "sausage," repurposed by geologists to describe a particular structure formed by rock deformation deep inside a mountain belt. A boudin is a detached lozenge of a once-continuous rock layer that has been stretched and locally thinned to the point of disconnection, in the manner that a piece of Silly Putty "necks" down when pulled apart. Typically, boudins form in sets as the layer is dismembered, creating a chain of intact lumps linked by highly deformed rock, a bit like sausages on a string.

Although they reflect significant degrees of deformation, boudins form only in relatively strong "competent" layers that are sandwiched between weaker, more ductile "incompetent" ones that can simply ooze or flow in response to applied stresses. The cross-sectional shapes of boudins are governed by the contrast in rock strength between the competent and incompetent layers. If the strength difference is small, the individual boudins will have a tapering lentil-like form, while if the difference is great, the boudins are chunkier: barrel- or brick-shaped.

In cases of extreme deformation, boudins can become completely detached from each other and "float" in the matrix of the weak enveloping rock, creating puzzling geometries that require some creative thinking by geologists.

I recall stopping at an outcrop on a field trip to the Adirondacks in which a beautiful white marble—metamorphosed limestone, formed on the seafloor—enclosed scattered toaster-sized cubes of black basalt, an igneous rock. At first glance, this seemed completely preposterous: How could a sedimentary rock from a quiet marine environment contain chunks—especially such large and oddly shaped ones—of magmatic rock? Normally it's the other way around: as magmas move up through the crust, they engulf bits of the host rock through which they ascend (see *Xenolith*).

After some minutes of head scratching and hypothesis pitching, our group had an epiphany: the basalt had started as a dike, or planar intrusion, into the limestone, emplaced along a fracture. Then, during the mountain-building event that formed the Adirondacks and transformed the limestone to marble, the strong, tabular basalt was broken into blocky boudins while the softer marble oozed into the spaces between them.

And after all that strenuous deductive reasoning, we were really hungry.

See also Breccia; Mylonite; Xenolith.

Breccia [BRETCH-ah]
You're breaking up
From the Italian for "breach" or "break," **breccia** is used in geology to describe a rock that is a mosaic of angular fragments. A rock with such a texture can form in several ways.

It could be a *sedimentary* breccia, with eroded chunks of rock that haven't traveled far enough to become rounded. Another possibility is a *tectonic* breccia, formed by shearing and grinding in a fault zone, or a *collapse* breccia, common in limestones that have dissolution cavities or other karst features (see *Karst*).

The most dramatic origin would be sudden shattering by a rogue meteorite—forming an *impact* breccia. In the impact scenario, there are further subtypes of breccia: *fall-back* breccia, formed when material that was ejected violently from the impact site rained back down into the resultant crater, and *in situ* breccia, which records the extreme compression and rapid decompression experienced by the target rocks. Most of the rocks brought back from the moon by Apollo astronauts are varieties of impact breccia. For earthly breccias, there are diagnostic clues—for example, a rock's setting, mineralogy, and range of fragment sizes—that allow geologists to determine the nature of the trauma that led to their broken-up character.

The term "breccia" highlights a nomenclatural challenge in geology: the perennial tension between purely descriptive names versus "genetic" ones based on inferred processes. Descriptive names like breccia can be vague, but they are potentially more durable than those linked to a particular mode of formation, because such terms may become obsolete as interpretations change.

But if you find a rock consisting mainly of shards and fragments, you know what to call it?

You breccia!

See also Karst; Mylonite; Pseudotachylyte; Twist Hackle.

Brimstone

Volcanic theology

Although primarily a metaphoric (not to be confused with metamorphic) rock, symbolizing the unpleasant prospect of eternal damnation, **brimstone** is an actual physical entity—namely, elemental sulfur in crystal

form. Its main settings of formation are stinky, sulfur dioxide–spewing magmatic vents, known as *solfataras*, after a particularly pungent one in the volcanic region that includes Mount Vesuvius, near Naples in Southern Italy.

Brimstone is not only not metamorphic but it also defies classification into either of the other two major categories of rock: igneous (cooled from magma) and sedimentary (deposited by water, wind, or ice). Instead, it is a sublimate—a material crystallized from a gaseous vapor directly into solid form. Its rotten-egg odor and association with divine wrath, however, make it hard to classify as sublime.

See also Amygdule; Lusi.

Chondrite [KON-drite]
Recipe for a small planet
Chondritic meteorites, or ***chondrites***, are samples of the raw ingredients from which the solar system formed. About 85 percent of witnessed meteorite falls and 67 percent of all meteorites found are chondrites. Their name comes from a Greek root meaning "grain," a reference to the distinctive, round, sand-sized particles called "chondrules" that give these meteorites a granular, almost sedimentary-looking texture. But these are sediments like no other. Older than Earth or any of the other rocky planets, chondrules represent "droplets" of rocky and metallic material that condensed at a high temperature from the primordial gas and dust cloud (the solar nebula) soon after enough mass had accumulated at its center to ignite the sun.

Through gravitational attraction, these solidified droplets began to stick together and accrete into larger bodies. The largest such masses had enough heat energy to melt, differentiate (separate) into metallic cores and silicate mantles, and become rocky planets. Iron meteorites and nonchondritic stony meteorites represent the cores and mantles, respectively, of ill-fated early planets that had differentiated but were then shattered by

collisions with other orbiting objects. In contrast, chondritic meteorites come from smaller "planetesimals" that never differentiated but instead remained homogenous samples of the primitive solar nebula material.

Chondrites—or materials with chondritic composition—were thus the universal ancestors of all rocks on Earth and other rocky planets and moons. They preserve a memory of the starting composition of these worlds, something that all other rocks have forgotten, particularly on Earth, where the process of differentiation has continued for 4.5 billion years and generated a prodigious diversity of rocks that fall far from the chondritic tree in their chemistry.

The crust has been distilled from the mantle by partial melting—a kind of smelting process that can yield basaltic ocean crust in one step but requires many iterations to generate granitic continental crust, unique to Earth. From the start, water (much of it delivered by comets) has interacted with both types of crust, inventing rocks and minerals never foreseen by the ancestral chondrites. Life got involved early on as well, creatively combining and recombining elements in novel ways, creating still other mineral varieties.

As products of all this differentiation, refining, and remixing, no native earth rocks can tell us about the bulk composition of the planet. To reconstruct that, we'd need to rehomogenize the earth—put it in some sort of colossal food processor and then sample the puree. Luckily, this isn't necessary, because chondrites—those ancient emissaries with astonishingly long memories—occasionally drop in from the sky and recall the recipe in detail.

See also Acasta Gneiss; Granitization; Kimberlite; Simplified Geologic Timescale (Appendix 1).

Cryogenian [kry-oh-GEN-ee-un]
Many are cold but few are frozen

In Norse mythology, life emerged from ice. In the beginning was the frost giant Ymir, whose cow liked to lick salty, rime-covered rocks, and in doing so one day uncovered Buri, the progenitor of the Norse gods, the Aesir. Ymir, representing the wild power of nature, was then killed by these more sophisticated beings, but he was not vanquished; his blood became the ocean, and his bones and teeth, stones.

I think of this strange and violent myth whenever I hear the term *Cryogenian*—derived from Greek, meaning "born from ice." The Cryogenian is a formal period-level division of the geologic timescale late in the Proterozoic eon of Precambrian time, from about 720 to 635 million years ago, more familiarly known as "Snowball Earth." Because it was such an extreme climate excursion—and seems to have changed the arc of the evolution of life on Earth—it merits a longer entry in these pages.

Rocks recording an ancient glacial period, just before the dawn of macroscopic animal life, were recognized as early as the 1880s in Scotland and Norway, but given the comparatively high latitudes of these places, the occurrence of rocks left by glacial ice did not require radical climate hypotheses to be proposed. And in many other parts of the world, there is no record of this period, which corresponds with the erosional gap called the "Great Unconformity" in the Grand Canyon.

Starting in the 1960s, however, geologists began to realize that something unusual had happened in the late Proterozoic time. Everywhere in the world that strata of this age do occur, they are "diamictites"— rocks consisting, improbably, of everything from fine clay to large boulders. This combination of grain sizes could not have been deposited by moving water, which leaves sediments neatly sorted by size as a function of velocity. Instead, diamictites are the signature of glacial ice, which carries motes of dust and mammoth-sized boulders with equal ease and leaves characteristically disorganized sediments behind. The global distribution of late Proterozoic diamictites suggested an unusually severe ice age, one far more extreme than the recent one in the Pleistocene, which ended only 10,000 years ago.

The plate-tectonic revolution of the mid-1960s brought the realization that rocks—especially those more than a few hundred million years old—may have moved far from the location on the globe at which they formed. But new methods for determining the paleolatitudes of rocks from their magnetic minerals had also been developed by this time, and these confirmed that some of the late Proterozoic glacial deposits had in fact been laid down in the tropics and even at the equator. Although the idea of an earth frozen from pole to pole ran counter to geologists' instinct for uniformitarianism, it was becoming clear that for some reason the planet's thermostat had gone haywire at the very end of the Precambrian.

Continued study of the rock record of this long, frigid spell, which geologists began to call Snowball Earth, has led to a more detailed understanding of its context and aftermath. The deep freeze was probably caused by the unusual clustering of most of Earth's land masses in a supercontinent (Rodinia, Russian for "motherland") that lay almost entirely at low latitudes. An ordinary episode of global cooling may have begun, and growing areas of sea ice near the poles would have caused increasing amounts of solar radiation to be reflected back to space. Meanwhile, another powerful cooling process, Earth's primary mechanism for removing volcanic carbon dioxide (CO_2) from the atmosphere, would have continued at pre–ice age rates: namely, weathering of continental rocks by rainwater carrying CO_2.

In a normal ice age, this natural carbon sequestration process is slowed when high-latitude continental areas begin to be covered by glacial ice. But in late Proterozoic time, with most of the land on Earth concentrated around the equator, weathering of continental rocks would have continued unabated, allowing sea ice, and later land-based ice caps, to expand until the world was a white, reflective snowball unable to warm itself. Most geologists agree that this was probably the most extreme climate event the planet ever experienced, but we continue to argue about whether the entire globe was truly frozen over for extended periods of time. That is, was Earth really a hard snowball or more like a giant slushy?

At this point in Earth's history, microbial life had been around for almost three billion years, and although organisms were small, they were diverse in their strategies for survival. Photosynthesis was a long-established habit, as was harvesting energy from chemical reactions at volcanic vents or from the waste products of other organisms. All of these ancient metabolic practices—and most notably photosynthesis, which requires access to sunlight—are still employed by organisms today. This indicates that somehow a wide range of life-forms survived Snowball Earth, and further suggests that at least some areas of the ocean remained open enough to allow photosynthesis to continue, perhaps in polynyas kept ice-free by upwelling currents.

Ultimately, a buildup of volcanic carbon dioxide in the atmosphere may have broken the world out of its icy state. While Earth's biosphere was not vanquished by the extreme and prolonged cold of Snowball Earth, it was profoundly changed by it. Rocks immediately above the last glacial deposits contain fossils of macroscopic organisms, the enigmatic Ediacara (derived from the Aboriginal language meaning "a veinlike spring of water"), orders of magnitude larger than any seen in pre-Snowball rocks. And soon after these pioneers, the ancestors of all modern animals burst onto the scene in what is known as the Cambrian explosion.

Why the period following Snowball Earth was a time of such biological novelty is another subject of lively debate. Sea level would have risen dramatically as glaciers melted, creating vast, sunlit continental shelf areas ideally suited for life. Organisms refilling these fertile ecological niches would have had unprecedented latitude to experiment. The end of the long ice age was also a period of radical changes in ocean chemistry: a well-documented jump in oxygen levels almost certainly catalyzed evolutionary innovation. Rock abrasion and erosion by Snowball glaciers not only contributed to the formation of the Great Unconformity but also delivered the gift of phosphorous to the seas, catalyzing a frenzy of biological productivity.

In 1990, the International Commission on Stratigraphy, or ICS, the body that governs the geologic timescale with slow and deliberative rigor, formally adopted the term "Cryogenian" for the Snowball Earth interval of the late Proterozoic. ("Snowballian" apparently just didn't have sufficient gravitas.) While the Phanerozoic eon—the time of "visible life," from the Cambrian to the present—has been divided into periods and finer intervals since the late 19th century, the Precambrian (representing 8/9 of Earth's history) lacked formal, named divisions until very recently. To geologists of the Victorian era, fossil-poor Precambrian rocks were largely inscrutable. But 150 years of geologic mapping and analysis, combined with powerful new dating techniques (particularly uranium-lead zircon dating), have made it possible to reconstruct in detail the chronology of events in Earth's first four billion years. So the ICS is finally giving names to the vast stretches of Proterozoic and Archean time that were previously referred to only by their isotopic ages.

There are two other geologic time units named for the rock types that are most characteristic of those intervals. The Carboniferous in the late Paleozoic era takes its title from the widespread coal deposits of that period, when the world's first great forests took root. And Cretaceous comes from the Latin for "chalk," a reference to the ubiquitous layers of white limestone laid down in a hothouse world with exceptionally high sea levels. "Cryogenian," then, satisfied the ICS's preference for precedent, having parallels, both alliterative and thematic, with these long-accepted names for intervals in the geologic past.

The Cryogenian also has odd parallels with the Norse story of Ymir and his cow, and how our world emerged from a kingdom of ice. Geologists increasingly see the Cryogenian as a pivotal moment in the planet's evolution, the event that gave rise to the modern biosphere. For an endless winter, frost giants reigned and rime-covered stones were scoured, until at last the realm of ice fell and the oceans swelled, creating a capacious world where new kinds of creatures could arise.

See also Ediacara; Polynya; Unconformity; Zircon; Simplified Geologic Timescale (Appendix 1).

Darcy [DAR-see]

A truth universally acknowledged

For the many fans of Jane Austen, there is only one Mr. Darcy: the dashing, aloof, and quintessentially English match for Elizabeth Bennet in *Pride and Prejudice*. But in the world of geology, there is another famous Mr. Darcy, a near contemporary of his fictional counterpart but from the opposite side of the Channel. Henri Philibert Gaspard Darcy (1803–58) was a French civil engineer who laid the groundwork for the modern study of groundwater. Fittingly, a fundamental quantity in hydrogeology, permeability, is measured in units bearing his name, the small-d ***darcy***.

Darcy's great achievement was designing an efficient public water system for the city of Dijon, which faced a crisis because its wells did not yield sufficient water for its population. Darcy's ingenious scheme conveyed water from a spring outside the city through a network of pipes and sand filters and was powered entirely by gravity, with no pumping required. In order to better modulate flow rates, he designed scale models of the system and formulated the first quantitative description of the movement of groundwater through sediments and rocks, now known as Darcy's law.

In simplified form, this equation states that the rate of groundwater flow through a given cross-sectional area of sand or other material is equal to the product of two factors. The first is the driving force: the fluid pressure gradient, or change in pressure over distance. Headstrong Elizabeth Bennet, insisting she must walk to see her ailing sister despite heavy rain, could just as well have been describing this component of Darcy's law when she said, "The distance is not great, if one has the motive."

The second factor governing how quickly groundwater flows is the permeability of the medium—the ease with which water can move through it—quantified by the unit named for Darcy. Muddy fields turn out to make Elizabeth's walk much slower than expected despite her strong urge to see her sister; she had overestimated both the literal capacity of mud to transmit water and its literary "permeability" to pedestrians.

Intriguingly, Darcy's law has the same basic form as flow laws for other entities, including Fourier's law for heat conduction and Ohm's law for electricity: the flow rate (heat flow or current) is equal to a potential gradient (temperature difference; voltage) times a constant describing how efficiently a material can convey the entity (thermal or electrical conductivity).

Permeability, the analog of conductivity in these other laws, varies over many orders of magnitude in geologic materials. Sand, the medium Darcy originally worked with, has a permeability of about one darcy, while a coarse gravel with large open spaces may be 10,000 times more permeable. At the other extreme, the permeability of clays or tightly crystalline igneous rocks is measured in milli- or microdarcys (one-thousandth and one-millionth of a darcy, respectively). And even within a given rock or sediment, permeability may be strongly anisotropic—varying significantly with direction—for example, much larger parallel with layering than perpendicular to it. It should be noted that Darcy's law applies only to seepage of groundwater through porous media—not to potentially turbulent flow through large voids or cave channels in karst terrains (see *Karst*), where the normal hydrogeological rules are suspended.

In earlier times, groundwater phenomena like seeps, springs, and artesian fountains were often considered enchanted or supernatural. Even today, elements of the occult linger in beliefs about groundwater; dowsers, water witches, and divining rods have all survived into the digital age. Darcy was among the first to demystify the hidden realm of groundwater. In rather the same way that Austen's inscrutable Mr. Darcy is gradually revealed to be reasonable and principled, Henri Darcy showed that groundwater follows rational physical laws, free from superstition, pride, or prejudice.

See also Karst; Speleothem.

Deborah Number
Beyond measure
Having standardized units of measure—liters, lumens, light-years—is essential for everything from cooking to space travel, but sometimes, paradoxically, the best way to size things up is with numbers that have no units. Geophysicists do this frequently as they grapple with fluid phenomena that occur over many temporal and spatial scales, from ocean currents to the overturn of the solid mantle.

One example of such a dimensionless measure is the **Deborah number**, named for the biblical priestess in the book of Judges who sang (at least in some translations), "the mountains flowed before the Lord." Like other geophysical metrics of this type, the Deborah number is a ratio, in this case of two time intervals. The numerator is the "relaxation time" of a medium—the time it takes for a fluid to reach an equilibrium state after it is loaded or unloaded; for example, how long it takes a handprint to disappear from a "memory foam" mattress or a landmass to rebound from the weight of a glacier. The denominator is the time of observation, which for actual flesh-and-blood humans is limited to decades but for computer simulations (and divine observers) can be millions of years or longer.

If the Deborah number is significantly greater than one, the material behaves as a solid on the timescale under consideration. If it is close to one, it is a fluid—not necessarily a liquid but rather something that flows. A further implication of the Deborah number is that depending on how long one cares to watch, a perfectly rigid, seemingly permanent thing—like a mountain belt—may in fact act like molasses over longer timescales.

A related quantity is the Rayleigh number, a metric for whether a fluid heated from below is likely to convect, or overturn, as a result of thermal expansion, applicable to entities ranging from lava lamps to planetary mantles. The numerator of the Rayleigh number includes all factors that favor convection, such as the force of gravity, the temperature difference from top to bottom, and the amount by which the material expands when heated. The denominator is a composite of variables that resist convection—especially the viscosity, or rigidity, of the medium, and its thermal conductivity (which, if high, will erase temperature differences). If the Rayleigh number is above a critical value, we can be confident that a given material, such as Earth's mantle, will convect, even if we cannot observe that behavior over human timescales.

Counterintuitively, abandoning our instinct to measure—forgoing furlongs, firkins, and foot-pounds in favor of these dimensionless ratios—can allow us to truly fathom nature, and may even provide something like a god's-eye view of the world.

See also Eclogite; Kimberlite.

Dreikanter [DRY-kahn-ter]
Three, of a kind

When our German-language class in high school was covering the conditional case, our teacher had us learn the song "Mein Hut der hat drei Ecken":

> *My hat it has three corners; Three corners has my hat.*
> *If it didn't have three corners, then it wouldn't be my hat.*

I think of this ditty and its tautological declaration whenever I hear the term *dreikanter*, a German word literally meaning "three edges," and used in geology to describe a rock that has been faceted by desert winds. Stones in the desert are subjected to constant sandblasting, and if they are hard and crystalline, they can develop polished, fluted, flat faces that reflect prevailing wind directions. NASA's Mars rovers have encountered many handsome dreikanter on their robotic rambles, attesting to the ferocity of winds on that planet.

A more general term for wind-sculpted rocks is "ventifact," and technically only pyramidal rocks with three sharp edges should be called "dreikanter." But many geologists apply the term elastically to stones with four or more edges, not as obsessed with threeness as the owner of the tricorn hat.

See also Erg; Yardang.

E clogite [ECK-loh-jite]
Pulling its weight

The metamorphic rock called *eclogite* could be likened to the nighttime workers who clean streets, schools, and offices: seldom seen but essential to keeping the world functioning smoothly. Eclogite is rarely found at Earth's surface, and even many geologists have never glimpsed it in outcrop. But without eclogite, the planet's plate-tectonic system would grind to a halt. As if that weren't compelling enough to merit our attention, eclogite is also an exquisitely beautiful jewel-tone rock with raspberry-red garnets set in a matrix of green and blue minerals.

Like all metamorphic rocks, eclogite has two distinct chapters in its biography: its birth in one environment and its subsequent transformation via recrystallization upon finding itself in different physical conditions. Most eclogites begin as the familiar black volcanic rock called "basalt," which is the most common rock type on Earth's surface, underlying all of the world's ocean basins.

Seafloor basalt is churned out at mid-ocean ridges, the globe-encircling chain of submerged mountains whose topography was first mapped by the great marine cartographer Marie Tharp of Columbia University in the late 1950s. Tharp laboriously compiled linear depth-sounding records from ship tracks into dramatic three-dimensional renderings that revealed the ruggedness of the deep ocean floor and forever dispelled notions that it was flat and featureless. Her mapping of the global mid-ocean ridge system was essential to the plate-tectonic revolution and discovery of the process by which continents could "drift": seafloor spreading.

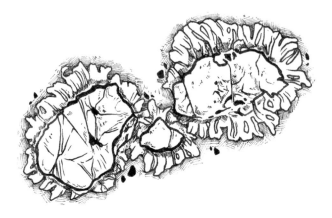

In seafloor spreading, basalt is erupted from fissures at mid-ocean ridges, where magmas are generated by partial melting of the uppermost mantle. These magmas are very different in composition from the mantle rocks from which they come—in much the same way that the first drops from a melting popsicle are richer in sugary juice than the popsicle as a whole. Specifically, basalt contains more silicon, aluminum, and calcium than the magnesium-rich mantle, which is dominated by the mineral olivine (whose gem form is peridot, the olive-green August birthstone).

After a given batch of basalt is emplaced, it is pushed progressively farther away from the submarine ridge as new generations of magma rise up and displace earlier ones. Following their fiery birth, ocean basalts spend the rest of their days quietly losing heat, until at the age of 150 million years or so, far from their place of origin, they become cool and dense enough to begin to sink back into the earth's mantle, in the crustal recycling process called "subduction."

But even very cold basalt is too buoyant to descend very far into the high-density olivine rocks of the mantle. If nothing further happened to it, old basaltic ocean crust would simply pile up in stacks at subduction zones, forming great mountain ranges of basalt rather than being returned to the mantle "forge." However, something remarkable does happen late in life to sea-floor basalt: when it reaches a depth of 30 miles or so into the mantle, it undergoes a profound metamorphic conversion. Its original aluminous minerals reconfigure themselves to much denser forms, and drab basalt is reincarnated, improbably, as technicolored eclogite with deep red garnets, grass-green pyroxene, and sky-blue kyanite. In this new guise, the rock that was once lighter than the surrounding mantle is now much heavier and able to sink deep into Earth's interior. The dense mass, in turn, pulls more unmetamorphosed ocean crust down to the depths where it, too, will convert to eclogite.

It's not an overstatement, then, to say that the formation of eclogite drives subduction, and because subduction is a signature process of Earth's tectonic system, the planet as we know it would be utterly different without eclogite. Although Venus, Mars, Mercury, and the moon show evidence of past volcanism and some crustal deformation, only Earth has developed the habit of subduction, which has helped to keep the planet on an even keel for eons. Subduction keeps the interior and exterior of the earth in communication with each other, returning not only the solid ocean crust but also volatiles like water and carbon dioxide, vented by volcanoes, back to the mantle. In contrast, other planets, like Mars with its fossil river valleys, have simply lost their volatiles to space over time, with nothing held in reserve.

On Earth, water carried down with a subducting slab locally reduces the melting temperature of the adjacent mantle rock, yielding granitic magmas that build continental crust over time. Water also lowers the viscosity of the solid mantle as a whole, allowing it to flow convectively and thereby keep the plates in motion. Current estimates are that the volume of water in the mantle is greater than that in the world's oceans—a comforting planetary nest egg that eclogite has set aside for us.

So let us sing the praises of eclogite. Although it generally stays out of the limelight, doing its work deep in the subsurface, very occasionally—by tectonic mechanisms not fully understood—some eclogite does find its way back to Earth's surface, where it can be rightfully admired and acknowledged for its indefatigable service to the planet.

See also Benioff-Wadati Zone; Granitization; Kimberlite.

Ediacara [ee-dee-ACK-ah-rah]
Peaceable kingdom
In the beginning, the earth was without forbs . . . or any other plants or animals. But soon after the primordial period of meteorite bombardment subsided around 3.8 billion years ago, microbial life emerged and gained a global foothold (even though feet were still far in the future). For at least 3 billion years after that, Earth's biosphere was a tranquil unicellular utopia.

Things might have continued like this indefinitely if the planet hadn't slipped into a long, deep ice age known as the Snowball Earth interval or, more formally, the Cryogenian period. Surviving this winter of winters would have required extraordinary toughness and versatility, and luckily some living beings were up to the task.

Remarkably, rocks that immediately overlie the last Snowball Earth glacial deposits preserve fossils not only of microbial communities but of macroscopic life-forms that were part of complex ecosystems. These mysterious organisms are known collectively as the ***Ediacara***, for the area in South Australia where they were first discovered in 1946. Ediacaran fossils have subsequently been found at forty other sites around the world, ranging from Norway to Namibia.

For a number of reasons, the Ediacara are paleontological enigmas. First, they seem to appear out of nowhere 635 million years ago, diverse and fully formed, with no clear antecedents. Although this can be attributed in part to the imperfection of the fossil record, the scant time between the end of the Cryogenian glaciation and the global appearance of this entirely new group of organisms is astonishing.

Their anatomical variety is also impressive. Some Ediacara, the so-called *rangeomorphs*, look like bloated ferns—frond-like and branching but inflated rather than flat. They towered over the seabed at more than three feet tall with holdfasts to anchor them in place. Other Ediacara were disc-shaped, including the radially grooved *Kimberella*, which appears to have been able to stretch to twice its size and shrink again, like an elastic Frisbee. The bizarre *Tribrachidium*, resembling a mini muffin-top, had a pattern on its crest with a threefold symmetry, seen in virtually no living group of organisms (but weirdly similar to the triskelion symbol on the flag of the Isle of Man).

And it's unclear what the Ediacara were made of. At several fossil sites, they occur as high-fidelity impressions in sand, which is not normally a good medium for preserving anatomical detail. This has led to the speculation that their exteriors were made of somewhat resistant yet still pliable material, perhaps an unknown biomolecule or finely crystalline, opaline silica.

The lifestyles of the Ediacara are equally mysterious. At the famous Ediacaran fossil site at Mistaken Point, Newfoundland, the organisms are preserved in life position, in a deep-sea turbidite deposit, indicating that they were living in darkness on the abyssal ocean floor, where photosynthesis would not have been possible. The fossils also lack any evidence of digestive systems—

and this, together with the large surface areas created by their ridged and branching forms, suggests that they somehow absorbed nutrients directly from seawater through a process called "osmotrophy." In other words, this was a world without predators, or even herbivores, a peaceable kingdom of strange, stretchy, puffy creatures that lived out their lives swaying placidly on the seafloor—the Garden of Ediacara.

But in later Ediacaran fossil strata, there are signs of unrest. Crawling tracks in the sediment indicate that organisms were on the prowl, some Kimberella specimens bear scratches that look like scars, and a growing proportion of the population has mineralized exoskeletons, suggesting that body armor was becoming a necessity. By early Cambrian time, 40 million years after they first appeared, the Ediacara vanished entirely—either driven to extinction by a new lineage of organisms with no qualms about eating fellow Earthlings or, perhaps in some cases, having themselves evolved into such organisms.

From that time onward, animal evolution has been an arms race between predators and prey. The first bite was the original sin.

See also Bioturbation; Cryogenian; Taphonomy; Turbidite; Simplified Geologic Timescale (Appendix 1).

Erg

The sands of time

To physicists, an erg is a tiny unit of energy, whose name comes from the Greek *ergon*, work; it's about the amount of effort needed to move a paper clip one-hundredth of an inch. To geologists, an **erg** is a large sea of sand, from the Arabic word meaning a "region of shifting dunes."

Measuring tens of square miles, ergs must have an ex-
tensive upwind source of sand—for example, an arid
floodplain or a beach—and a downwind barrier that
keeps the collected sand confined. Like an actual sea
whose surface is wrinkled by waves, ergs are rucked by
dunes of various sizes that march slowly across the landscape at the dictate of the prevailing winds.

In an inversion of the geologic mantra of uniformitarianism, "the present is the key to the past," ancient
ergs preserved as sandstone can shed light on how
modern dunes migrate over the landscape, something
that is difficult to observe in three dimensions in real
time. In the steep canyon walls of Zion National Park,
for example, where the lower Jurassic Navajo Sandstone
is spectacularly exposed, one is able to see the internal structure of massive sand dunes formed 180 million years ago. The sweeping "cross beds"—strata inclined to the horizontal, so characteristic of the rocks
in Zion—represent the downwind slip faces of ancient
dunes that were continuously on the move, climbing
up the backs of other dunes as they themselves were

overridden. The sandstones at Zion are in effect an immense Jurassic weather almanac from which it is possible to reconstruct wind patterns and rainfall cycles at the time of the first dinosaurs.

It's dizzying to contemplate the immense energy embodied by such an erg complex: the weathering and erosion to grind rocks into so much sand, the wind power to gather it in one place and then propel the great dunes across the desert. The ergs of physics simply aren't the measure of their giant geological counterparts.

See also Dreikanter; Katabatic Winds; Uniformitarianism; Yardang.

Firn [fern]
Snows of yesteryear

A Swiss German word meaning "from last year," *firn* is sugary snow that has survived for several seasons and is on its way to becoming glacial ice. As it is buried beneath the snows of successive winters, firn gradually compacts, and open spaces between the grains of icy snow slowly close. Once firn reaches depths of 200 feet or so, the old snow is fully recrystallized, and any remaining air that was intermingled with the snowflakes when they fell becomes trapped as bubbles in solid ice.

These tiny bubbles—natural vials of ancient air—provide a continuous archive of atmospheric composition for the past few hundreds of thousands of years. Air bubbles in ice cores from northern Greenland, where yearly snowfall is abundant, represent a very high-fidelity record of atmospheric gases going back to the previous interglacial period 120,000 years ago. Cores from Dome C station in Antarctica, where the snow accumulates more slowly, document atmospheric changes at lower resolution but extend backward through seven glacial full cycles to 800,000 years. Though literally poles apart, the pockets of old air from the far north and far south tell the same sobering story: that global temperature variations closely track the concentration of greenhouse gases, especially carbon dioxide and methane, and concentrations of those gases are higher today than at any time the ice remembers.

The snowflakes of yesteryear are now crystal balls foretelling our future.

See also Sverdrup; Varves.

Gastrolith
From the gizzards of lizards

A geologic neologism from the Greek words for "stomach" and "stone," a ***gastrolith*** is a fossil gizzard stone from a sauropod dinosaur or swimming reptile like a plesiosaur. Like modern poultry and some other plant-eating animals that lack grinding teeth, these voracious Mesozoic herbivores were actually also rock-eaters (petrovores?) who ingested stones to help grind up tough vegetation in their gut. But in contrast to chickens and turkeys, who swallow sand-sized particles, these dinosaurs and their marine cousins gobbled up cobbles as large as four inches in diameter. Leafy flowering plants (angiosperms), which today dominate the forests and grasslands, had not yet evolved in Jurassic time, and even the absurdly gigantic titanosaurs had to sustain themselves on the spiky, waxy needles and other prickly parts of conifers. Who could blame them for occasionally chowing down on chert or bolting some basalt?

Gastroliths can be easily identified as such if they occur together with dinosaur bones, but even if their former hosts didn't make it into the fossil record, gastroliths can be recognized on the basis of several criteria. First, they must be out of place geologically—for example, an anomalous pebble, or a cluster of them, embedded in an otherwise fine-grained rock like shale or limestone. Second, gastroliths tend to be remarkably smooth and polished, as if processed in a rock tumbler—which in a sense they were, though for the squeamish it may be best not to contemplate this image in too much detail.

See also Geophagy; Taphonomy.

Geodynamo [gee-oh-DIE-na-moh]

Electromagnetic blanket

Invisible but essential, the Earth's magnetic field is perhaps the least celebrated of the many environmental amenities provided by this hospitable planet. It is also one of the strangest. It has existed for more than 3.5 billion years but fluctuates daily. It emanates from Earth's deep interior but extends far out into space. It is intangible and mostly unseen—except when it lights up in ostentatious greens and reds during the auroras—but crucial to life. The magnetic field is our protective shield; it deflects not only the relentless solar wind, which could otherwise strip away Earth's atmosphere over time, but also cosmic rays, which zing in from interstellar space with enough energy to damage living cells.

Although sailors have navigated by the magnetic field for a millennium, and scientists have monitored it since the 1830s, much about it remains an enigma. Einstein himself said that understanding its origin and longevity was one of the great unsolved problems in physics. Although aspects of the magnetic field remain mysterious, the scientific consensus today is that it arises in Earth's outer core, where the movement of liquid iron creates a giant, self-perpetuating electromagnet, the *geodynamo*.

The fact that the geometry of the field is approximately dipolar—like a bar magnet—and that the magnetic poles coincide, on average, with the geographic North and South Poles, indicates that Earth's rotation largely governs the motion of molten iron in the core. Until the mid-1930s, it was believed that Earth's core

NORTH POLE ~~IS~~ WAS HERE

was entirely molten. This was based on the observation that following a major earthquake, seismic S- (shear) waves, which cannot travel through liquids, are not detected on the opposite side of the globe. The paths of P- (pressure) waves, which are slowed but not stopped by liquids, are also altered in a manner that suggested the core, with a radius of about 2,100 miles, was molten top to bottom.

But in 1936, the Danish geophysicist Inge Lehman—one of the few women in a field that is still overwhelmingly male—analyzed seismic records from a great earthquake in New Zealand and noticed a few distinct P-wave arrivals that were not consistent with the all-molten core hypothesis. She correctly interpreted these seismic signals as waves that had ricocheted against a crystalline mass deep in the earth—a solid inner core 750 miles in radius. Modern geophysicists now recognize that a significant component of the dipolar magnetic field is generated by differential rotation between the fast-spinning inner core and comparatively sluggish outer core.

However, Earth's pirouetting on its rotation axis can't explain the full complexity of Earth's magnetic field, which in addition to the dipole also has "quadrupole" and "octopole" components, making its actual geometry something like a playground jack with extra spikes. These components are thought to reflect more complex choreographic movements caused by thermal convection in the outer core.

For reasons still not well understood, the superimposed spinning and roiling motions of liquid iron in the outer core intermittently cause the polarity of the magnetic field to flip—with magnetic north and south poles changing places over the course of a few millennia, during which the overall field strength is significantly lower. Seafloor basalts provide a high-fidelity chronicle of these reversals for the past 170 million years (at which point the record is lost to subduction).

Reassuringly, there is no clear paleontological evidence that magnetic reversals are linked with mass extinctions (though one wonders how migrating animals manage during the thousands of years it takes for a new magnetic regime to become established). The most recent flip, known as the Matuyama-Brunhes reversal, occurred about 770,000 years ago, deep in the Ice Age. Our Pleistocene ancestors likely never noticed, but we Anthropocene humans surely would: such an event would cause debilitating disruptions to the electrical grid and satellite communications.

So is the behavior of the magnetic field yet another planetary matter that should keep us awake at night? Since 1990, the Magnetic North Pole has migrated almost 900 miles, from Axel Heiberg Island in the high Canadian Arctic to a site close to the true Geographic North Pole, and the intensity of the field has been falling at a rate of about 6 percent per century. It's frankly hard to know whether this is cause for alarm. In my view, we're better off worrying about the much greater likelihood of catastrophic climate change—which we can do something about—than about the elusive, magnificent magnetic field, which is completely beyond our control, but to which we should raise a nightly prayer of thanks.

See also Areology; Chondrite; Mohorovičić; Simplified Geologic Timescale (Appendix 1).

Geophagy [gee-OFF-ah-gee]

Raw terroir

In some cultures around the world, humans eat soil or pulverized rock, especially clay and chalk, in a practice called **geophagy**. Although this is sometimes associated with times of famine, anthropologists disagree about the reason for the habit. The nutritive "minerals" (actually elements) that we require—for example, calcium, sodium, iron, and zinc—generally come to us via plants, but plants in turn assimilate them from the soil. Sommeliers claim to be able to taste the distinctive mineral composition of soils in which wine grapes are grown. Geophagy is a more efficient (if less epicurean) way of sampling the goût de terroir.

See also Gastrolith; Pedogenesis.

Geosyncline [gee-oh-SIN-kline]

Magic mountains★

Mountains have long posed physical challenges to humans; their rugged terrain and capricious weather can be perilous for the unprepared. Mountains have also presented steep intellectual challenges for generations of geologists who did not have the theoretical "equipment" to scale them. Before the plate-tectonics revolution in the early 1960s, there was no good scientific explanation for the formation of mountains—and yet, in the words of George Mallory, they were there. For more than a century, an elaborate academic fantasy, the theory of ***geosynclines***, was the reigning paradigm for mountain growth. Now an embarrassing footnote in the history of geology, geosynclinal theory is perhaps best explained as the tectonic analog of spontaneous generation: like mice materializing in a sack of grain, mountains would spring up miraculously given the right ingredients.

★ Adapted from https://aeon.co/essays/when-geology-left-solid-ground-how-mountains-came-to-be.

As irrational as the theory may now seem, it was grounded in an accurate observation about North American geology: that as one follows rock strata eastward from midcontinent into the Appalachians, their thicknesses grow by an order of magnitude. In the Midwest, where the strata are flat-lying, they are thin; while in the mountains, where they are folded and contorted, they are at least ten times thicker. It was hard to imagine that this was mere coincidence.

In 1857 James Hall (1811–98), the state paleontologist of New York and founding president of the Geological Society of America, first articulated the idea that there could be a causal relationship between thick piles of sediment and the formation of mountain belts. Hall declined to explain in any detail why sedimentary basins would necessarily fold themselves, but he alluded to gravitational instabilities within a shivering mass of watery clay, silt, and sand. Hall's idea was adopted and expanded by James Dwight Dana of Yale (1813–95), a towering figure in 19th-century geology and author of the definitive encyclopedia on minerals, *Dana's System of Mineralogy*, a version of which is still in print today.

Dana coined the term "geosyncline" for a deep trough of sediment that (somehow) became the crumpled strata in mountain belts. The term "syncline" was already in use for U-shaped folds in rocks, and its antonym "anticline," for arch-shaped folds. Dana's neologism suggested down-warping on a grand scale that could encompass the smaller buckles and crinkles within a mountain belt. Such internal features were being described in increasing detail by practitioners of a new subdiscipline called "structural geology," which focused on modes of rock deformation. In the Appalachians, Alps, Scottish Highlands, and Canadian Rockies, hardy structural geologists who were mapping folds and faults began to converge on the same conclusion: that the rocks in all of these mountain belts had experienced significant amounts of horizontal contraction—squeezed to about 50 percent of their original lateral extent.

Horizontal telescoping of this magnitude was an uncomfortable fit with geosynclinal theory, which suggested that gravity was the primary driving force for mountain building. Perhaps to avoid painful cognitive dissonance (and denouncement by powerful figures like Dana), most field geologists of the time seem to have excused themselves from theorizing about causes, preferring simply to document what the rocks had to say.

A late 19th-century European school of structural geologists, however, proposed an explanation that seemed to reconcile the geosynclinal concept of gravitationally foundering basins with field-based evidence for horizontal shortening: that Earth's crust was shrinking as a result of cooling. In this view, mountain ranges were like the wrinkles on a raisin, and ocean basins were the downward involutions. The idea of a cooling, shriveling Earth was compelling in part because it brought geology into alignment with the cutting-edge science of the day: Lord Kelvin's thermodynamics.

For decades, Kelvin had been the bane of geologists, including Darwin, with his decrees that Earth was no older than 40 million years, based on arguments involving its present-day heat flow and intimidating calculations that few geologists could understand or refute. But by the 1890s, most geologists had resigned themselves to a foreshortened timescale and had internalized the idea that Earth was growing inexorably colder. (Kelvin did not know that Earth actually generates heat through radioactive decay—the same phenomenon that provides a means for dating rocks, and that eventually showed his pronouncements of Earth's age to be too young by a factor of 100.)

In 1922, the "Shrinking Earth hypothesis" acquired renewed credibility for another generation of geologists when the German geologist Hans Stille (1876–1966) published his opus, *Die Schrumpfung der Erde*, a grand synthesis of thermal contractionism with geosynclinal theory. In *Schrumpfung* and subsequent work, Stille introduced a scholarly taxonomy of geosynclines that seemed to elevate their study through Linnaean-like classification but in fact created chimeras that geologists would chase unproductively for decades. There were "orthogeosynclines," the proper ones, made of rocks that could be folded and that usually consisted of two subparts: a "eugeosyncline" with volcanic rocks and a "miogeosyncline" without them (today recognized as deep and shallow ocean sequences, respectively; eugeo-synclines included *ophiolites* and *turbidites*).

As late as 1958, the lexicon of geosynclines was still expanding. Marshall Kay, an eminent stratigrapher from Columbia University, published the third edition of his masterwork, *North American Geosynclines*. Kay's coinages reflect a particularly impressive command of Greek: "taphrogeosynclines" were fault-bounded; "zeugogeosynclines" formed on formerly stable continental crust; and "paraliageosynclines" were shallow and coastal, marginal to the real thing.

Some contemporary geologists grumbled about the absurd proliferation of prefixes, and many had privately lost their faith in geosynclinist doctrine, but few, particularly in the United States, were ready to cry publicly that the esteemed professors had no clothes. And then, less than a decade after Kay's treatise was reissued, geosynclinal theory was abruptly dethroned. Plate tectonics at last provided the motive force for building mountains: as the seafloor spreads, continents move and occasionally collide, crumpling the rocks on their margins.

The seminal observation that sedimentary sequences in mountains tend to be thicker than in continental interiors was accurate, but this is simply because sediments accumulate on continental shelves, which are the leading edge when continents collide. As Peter Coney, an American structural geologist, commented wryly in 1970: "Saying geosynclines lead to orogeny [mountain building] is like saying fenders lead to automobile accidents."

It is easy in retrospect to lampoon ideas that now seem so clearly misguided, but the geologists who ad-

vanced geosynclinal theory were motivated by the New-
tonian scientific instinct to find patterns in nature and
from these extrapolate universal laws. They correctly
identified an attribute common to many mountain belts,
but then, too eager to reach the summit, mounted the
intellectual equivalent of a Himalayan expedition without any of the proper scientific gear.

See also Allochthon; Klippe; Ophiolite; Turbidite.

Granitization
Igneous agnostics

Earth is the granite planet. Although its bulk composition is comparable to that of the other rocky planets (and
chondrite meteorites), only Earth has distilled significant
amounts of granite—the foundation for the continents—
from its mantle. So the question of the origin of granite
is fundamental to understanding how the planet works,
and the story of how geologic thinking about granite has
evolved is a fascinating glimpse of science at its best—
and worst.

The concept of ***granitization***—the misbegotten idea
that granites are not from melts but instead created
through the transformation of sedimentary rocks by in-
filtrating elements—was embraced between about 1930
and 1960 by geologists who felt they were at the fore of
an intellectual revolution, overthrowing the stodgy old
magmatic paradigm. Like the theory of geosynclines,
granitization began with some reasonable speculations
about a fundamental geological question but then mu-
tated into a mass hallucination that clouded geologic
thinking for decades. Granitization theory is rarely men-
tioned in geologic textbooks today, perhaps because it

doesn't follow the favored narrative of continuous progress in science, and many modern geologists aren't even aware of it.

To early geologists, long before the granitization delirium, the origin of granite was a puzzle. In the late 18th century, different rock types were thought to have formed during distinct periods in the geologic past. Granites, which often occur as the "basement" beneath stratified rock sequences, were called "primary" rocks and believed to record Earth's earliest eons. A school of thought known as "Neptunism"—led by the redoubtable German mining geologist Abraham Gottlieb Werner (1747–1817)—held that all rocks (other than those that clearly emanated from volcanoes) had formed by deposition in water. The Neptunists considered volcanism to be a minor, shallow phenomenon caused by burning coal seams, wholly unrelated to granites, which they interpreted as deposits from primordial oceans of alien composition.

At this same time, there were others, notably James Hutton (1726–97), the Scottish polymath credited with discovering Deep Time (see *Unconformity*), who recognized evidence that granitic rocks were not marine sediments but instead solidified melts. Hutton was, of course, right, though he was not exactly a paragon for objective, dispassionate science on this matter. He had previously formulated a grand theory—a prescient glimmer of plate tectonics—about how the earth's crust was endlessly churned and reforged by an internal heat source, and he was looking for corroboration of his scheme. (This is another embarrassing little fact rarely shared with students of geology, who are trained to make observations first, *then* draw inferences.)

Hutton's ideas have come down to us mainly as transcribed by his friend John Playfair (1748–1819), in his 1802 volume *Illustrations of the Huttonian Theory of the Earth*, a heartfelt tribute published five years after Hutton's death. In one of the book's most vivid excerpts, Playfair describes an outing in the Cairngorm Mountains, where a giddy Hutton found confirmation of his heat engine hypothesis. In the bed of the River Tilt, he came across fingers of pink granite that had made inroads into dark metasedimentary rocks in a manner that could not be explained unless the granitic material had been molten at the time. This crosscutting relationship also showed that granites were not invariably primordial but rather in some cases younger than other rocks. Playfair describes the eureka moment:

> In the bed of the river, many veins of red granite . . . were seen traversing the black micaceous schistus. . . . The sight of objects which verified at once so many important conclusions in [Hutton's] system, filled him with delight; and as his feelings, on such occasions, were always strongly expressed, the guides who accompanied him were convinced that it must be nothing less than the discovery of a vein of silver or gold, that could call forth such strong marks of joy and exultation.

We get a glimpse here of Hutton as a single-minded zealot, but the fact remains that he got very close to the truth in a way that many geological ideologues before and after him did not.

Over the course of the 19th century, geology matured from a pastime practiced by self-taught amateurs like Hutton into a profession with academic hierarchies and a growing range of analytical methods at hand. A special type of optical microscope, in which light was transmitted through thin slices of rock, allowed the mineral composition of granite and other igneous rocks to be determined in detail. Chemist Robert Bunsen, of the eponymous burner, demonstrated that the texture of granite —the nature of boundaries between crystals of different minerals—was consistent with cooling from a melt. By 1900, Hutton's intuitive concept of granite as a solidified magma seemed well established as empirical fact.

But then, beginning in the 1930s, there were rumblings of dissent. As laboratory methods improved, a burgeoning number of igneous rock types were identified, raising the question of how there could be so many distinct magma sources in Earth's interior: Was the mantle a "plum pudding" with pockets of different composition? Meanwhile, field geologists began to point out a geometric puzzle that became known as the "room problem": If large granite bodies like those in the Sierra Nevada had been intruded into other rocks, how had room been made for them—and what had become of the rocks they intruded? To some geologists, the obvious answer to both conundrums was that granites were not in fact magmatic but formed in situ by transformation of sedimentary rocks in a process they called "granitization."

This radical new view began with Helge Backlund, a Swedish professor from Uppsala, who had studied the odd *rapakivi* granites of the Baltic region. He argued

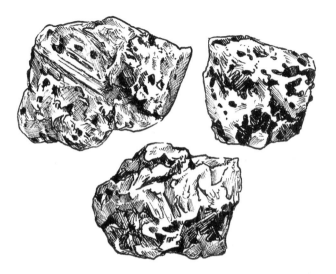

that the strangely rounded grains of the rapakivi granites, and what he perceived as layering within them,
were consistent with their having formed from stratified sedimentary rocks that were later altered via chemical "metasomatism"—something akin to the atom-by-
atom replacement by which buried tree trunks might
become petrified wood. Soon geologists in Britain
and the United States were arguing that they, too, had
found outcrops where sedimentary rocks graded im-
perceptibly into granites. Like a religion splintering
into sects, these "granitizers" or "transformists" then
began to break into subgroups that held different views
on the precise nature of metasomatism: Was it "dry,"
driven by heat and chemical gradients, or "wet," a consequence of deep crustal fluid flow? It's painful now to

read the hairsplitting sniping in the academic literature among geologists who were all studying the mane of a unicorn.

Even before granitization began to gain followers, a geochemist at the Geophysical Laboratory of the Carnegie Institution in Washington, DC, Norman Bowen, had in fact largely solved the granite problem. Beginning in the 1910s, Bowen created batches of magma in the lab from rocks of different compositions, then let them cool and crystallize. He showed that different minerals crystallize at different temperatures, and that if early formed minerals are removed from the remaining magma (in nature, via gravitational settling), then the residual melt will evolve toward compositions far removed from the original.

More specifically, Bowen showed that this process of "fractional crystallization" could yield small amounts of granite from an original melt with the composition of Earth's mantle. (This is conceptually similar to the process of fractional distillation at oil refineries, in which various types of hydrocarbons are separated from crude oil through their different boiling points.) If the full spectrum of igneous rock types could be generated by such a process, there was no need to invoke a mottled mantle of heterogeneous composition.

Bowen's brilliant work was largely ignored by the granitization crowd, who tended to be field geologists with deep suspicions about laboratory research based in the methods of physics and chemistry (a lingering result of earlier skirmishes with outsiders like the physicist Lord Kelvin, who scornfully dismissed geologists' well-founded arguments that Earth was hundreds of millions to billions of years old). Also, Bowen's results did not address the "room problem," which loomed large in the minds of the granitizers (the resolution, by the way, is that the "host" rocks are domed up by, and, to a lesser degree, assimilated into, intruding granitic melts).

By the late 1950s, however, with no progress toward elucidating the shadowy phenomenon of metasomatism, granitization theory was in decline. It was quietly abandoned in the 1960s when the paradigm of plate tectonics at last provided a framework for where and why rocks melt. Since then, geologists have also attained a subtler view of how Bowen's results actually apply in nature: It's unlikely that granite could be extracted in a single step directly from the mantle. Instead, fractional melting of already "refined" rocks, rather than fractional crystallization of a "primitive" mantle magma, is the likely genesis of most granites. And to be fair, some granites are derived from sedimentary rocks—but through melting rather than the mysterious processes of transformation advocated by the granitizers.

The story of how geologists have come to understand granite does not follow a simple, triumphant arc. Hutton's remarkable intuition proved correct, even though he violated modern protocols for the scientific method. Geologists who understandably believed that the truth could only be found in the rocks at the outcrop ended up becoming hopelessly lost in a fiction of their own making. Bowen demonstrated the power of bringing experiment into the discipline but was not interested in the messy complexities of granite in its natural habitats.

The lesson for geologists-in-training is that it takes a combination of fieldwork, theory, experiment, and imagination, all tempered by intellectual honesty, to investigate Earth, this grand old planet of granite.

See also Chondrite; Geosyncline; Rapakivi; Unconformity.

Grus [groos]

Things fall apart

Although granite is considered a symbol of strength and durability, and can survive billions of years deep underground, it is ultimately no match for raindrops and microbes at Earth's surface. **Grus**, from the Danish-Norwegian word for "gravel," is used in geology to describe thoroughly "rotten" granite—rock that has been so deeply weathered that it falls to pieces.

Grus is formed when water, often bearing organic acids, insinuates itself into granite along the boundaries of crystals and alters minerals through dissolution and hydration. Potassium feldspar, the mineral that gives granites their rosy pink blush, slowly changes to soft clays like kaolin. The black crystals of hornblende that pepper most granites turn into greenish flakes of chlorite. The interlocking mesh of crystals that formed as the granite solidified from a magma becomes weakened, the edges of grains no longer dovetailing. Rapakivi granites, with their unusual, rounded crystals, are particularly susceptible to this type of insidious disintegration.

Grus represents the natural first step in soil formation and can also be a valuable commercial commodity, sold for upscale driveways, landscaped paths, and some "clay" tennis courts. But grus is also a cautionary illustration of how quickly things that seem so solid—whether granites or the cultural bedrock of society—can crumble when components are compromised, connections are broken, and cohesion is lost.

See also Granitization; Pedogenesis; Rapakivi; Scree.

Haboob [huh-BOOB]
Dust in the wind

Like erg, *haboob* is an Arabic word for an impressive wind-related phenomenon—in this case, a looming wall of sand and dust propelled by a downdraft of dense air in a dry, sparsely vegetated region, often in advance of an approaching thunderstorm. Traveling at speeds up to 50 miles per hour, the leading edge of a haboob can be nearly a mile high—a menacing sight to witness, and a serious hazard for aircraft, drivers, and those with respiratory problems.

The term, which comes from the Arabic *habb*, "to blow," originated in the deserts of Sudan, where meteorological conditions conspire to produce dozens of haboob events each year. The adoption of the word into the geoscience lexicon can be traced to a 1972 paper in the *Bulletin of the American Meteorological Society*, which argued that the same processes that generate haboobs in Sudan also occur in Arizona.

Since the 1970s, the term has been applied to similar dust storms in a growing number of places around the globe—regions that never needed such a word, until unsustainable land-use practices and climate change accelerated desertification and began attracting haboobs to new territory.

See also Erg; Katabatic Wind; Yardang.

Hoodoo

Hats off

Like something from the sketchbook of Dr. Seuss, ***hoodoos*** are spindly, whimsical, gravity-defying towers of weathered rock that often occur in groups. Also called "goblins" or "fairy chimneys," many of them suggest either humanoid forms or unseen sculptors. The term "hoodoo" is probably a reference to their odd, otherworldly appearance.

Hoodoos form in horizontally layered rocks with vertical fractures that become enlarged over time by running water or by ice through the process of frost-wedging. The top of a hoodoo is always a relatively resistant rock that protects the softer underlying rock, temporarily, from erosion.

Depending on the particular rock type and the primary agents of weathering, these "caprocks" can have forms resembling headwear of diverse sorts: pointy wizard hats (Cappadocia, Turkey), tall fezzes (Bryce Canyon National Park, Utah), overhanging turbans (Drumheller, Alberta), or jauntily tilted fedoras (Putangirua, New Zealand). Once these caps are doffed by gravity or erosion—as if to say "Howdy Do"—the rest of the hoodoo will quickly succumb as well.

See also Grus; Pedogenesis; Yardang.

J ökulhlaup [YUH-kull-loip, very approximately, with aspirated Ls]
Breaking the ice

Hyperbole has no place in Iceland, where natural forces are intrinsically extreme. *Jökulhlaup* is an Icelandic word meaning "glacier run" (*hlaup* shares a root with the English words "lope" and "lap"). But don't be misled by the typically Icelandic understatement; a jökulhlaup is in fact a violent, high-volume flood caused either by the abrupt failure of an ice-dammed lake or by the sudden melting of glacial ice from volcanic heat. In other parts of the world, neither of these scenarios is likely to keep many people awake at night. In Iceland, both happen frequently enough that the government publishes detailed jökulhlaup preparedness plans.

The most recent major jökulhlaup in Iceland occurred in November 1996, when the volcanic vent called
Grímsvötn erupted beneath the country's largest glacier,
Vatnajökull. No one was particularly surprised, since
there are records starting from the 12th century that
Grímsvötn has been doing this regularly. Still, it was
inconvenient. The floodwaters, carrying 1,000-ton ice-
bergs, raged for a week and tore out 20 miles of the Ring
Road, Iceland's main highway. Several major bridges,
together with 100 million tons of ash and sediment, were
washed out into the North Atlantic. The indomitable
Icelanders rebuilt everything within a year—knowing
full well that the same thing will inevitably happen again.

The torrents unleashed by Grímsvötn, however, are
trickles compared with a series of gargantuan jökulh-

laups that occurred between about 18,000 and 13,000 years ago in the northwestern United States, forming an area known as the Channeled Scablands in eastern Washington State. The evidence for these floods was recognized in the 1920s by an iconoclastic geologist on the faculty at the University of Chicago, J. Harlen Bretz. Bretz could see from the ground level what would not be evident to others until aerial photographs became available: that the peculiar, rough landscape of the Channeled Scablands was essentially like the bed of a river of unimaginable size, with ripples spaced hundreds of yards apart, "gravel" the size of trucks, and potholes that could swallow entire farmsteads.

But at the time Bretz was mapping the Scablands, the orthodoxy in geology was a strict form of uniformitarianism that refused to accept any catastrophic explanations for geologic phenomena. The aversion to giant floods in particular ran deep in geologists' veins, owing to ongoing battles with biblical literalists who attributed modern landscape features—including the Grand Canyon—to Noah's flood. Bretz's hypothesis about a massive ancient torrent, orders of magnitude larger than anything in recorded history, seemed likely to open the floodgates (so to speak) to a new generation of creationists eager to demolish the whole edifice of geology. (The Scopes Monkey Trial, in which a Tennessee science teacher was charged with the "crime" of teaching evolution, happened during the years that Bretz was trying to make his case for the Scablands megaflood.)

A more serious scientific weakness in Bretz's hypothesis was that he could not point to a source for the biblical volumes of water that allegedly scoured the

Scablands. Bretz was publicly renounced by fellow geologists (and they were all fellows at that time) when he
dared present his "Spokane Flood" idea at professional
meetings. One of the few contemporary geologists who
dared to ally himself with Bretz was J. T. Pardee of the
US Geological Survey. Early in his career, Pardee had
documented the extent of a vast ice-age lake in Montana, Glacial Lake Missoula, and realized that this could
have held enough water to account for the outsized
flood features of the Scablands.

Over the following decades, using field observations, air photos, and fluid dynamic calculations, Bretz
and Pardee pieced together the story of the Scablands
floods. A lobe of the ice sheet that covered Western
Canada formed an ice dam that blocked the Clark Fork
River in what is now Missoula, Montana. Over time,
a volume of water equivalent to that in Lake Michi-
gan became impounded behind the ice until finally the
pressure was too great—and in less than a week, the
entire lake drained, in a flood that would surely have
impressed Noah.

The geological community gradually, grudgingly ad-
mitted that Bretz and Pardee were right—and, in fact,
subsequent workers have shown that there were as many
as 30 megafloods over 5,000 years as the ice repeat-
edly advanced, dammed the river, and then failed again
catastrophically. At its annual meetings, the Geological
Society of America (GSA) now holds symposia named
for Pardee, intended to be "safe spaces" where scientists
with ideas that challenge prevailing paradigms can speak
without fear of censure. And Bretz, who lived to see his
work venerated by a new generation of geoscientists,

received the GSA's Penrose Medal, its highest honor, in 1979, a few years before he died at age 99.

Many other examples of late ice-age jökulhlaups have now been recognized around the world. These violent events reflect the precarious instability of the hydrologic cycle as Earth made the transition into the Holocene. The Wisconsin Dells—best known as a kitschy tourist destination with arcades and water parks—first drew visitors because of its strange armada of oddly shaped sandstone islands along the Wisconsin River. These are now thought to have formed in a matter of days 14,000 years ago when an ice dam was breached and a glacial lake with 20 trillion gallons of water roared through the narrow stretch of the river, churning away at the sandstones in a hellish kind of premonition of the modern water parks.

New high-resolution bathymetry of the English Channel—the ditch that has dictated so much of modern European history—suggests that it, too, may be the product of a jökulhlaup, one to rival even those in the Channeled Scablands. For most of the Ice Age, with much of Earth's water locked up in glaciers, sea level was far lower, and it was possible to stroll from France into England, as the bones of many mammoths snagged by Channel trawlers attest.

But as glaciers retreated northward, a vast pool of water grew off the coast of the Netherlands, fed from the south by the Rhine and other rivers, and buttressed on the north by the still formidable ice sheet in Scandinavia. When that ice also began to melt, the lake was already brimful, and the added meltwater caused it to burst through its rim at what is now the strait between

Dover and Calais. As in the Scablands and the Wisconsin Dells, the entire volume of impounded water may have drained in less than a week, gouging out a submarine canyon called Hurd's Deep and forever severing the connection between Britain and the Continent.

One could argue that the Norman Conquest, the British Empire, Shakespeare, Winston Churchill, the Beatles—perhaps the English language itself—might never have come to be except for a jökulhlaup. And that's not even hyperbole.

See also Areology; Lahar; Sverdrup; Uniformitarianism.

Karst
The holey land

Borrowed from the name of a region in Slovenia famously pitted with sinkholes and caves, **karst** is used in geology as both a noun and an adjective to describe areas where limestone bedrock has been transformed into a rocky sponge through dissolution by groundwater.

Karst regions are like hydrological anarchies where the normal rules for surface and groundwater flow are simply ignored. Streams flowing along merrily at the surface suddenly disappear into subterranean channels. Groundwater, which typically filters through normal rocks at a few feet per year, may rush through interconnected cavities in karst regions at hundreds of feet per *day*. This hydrological free-for-all leads to easy contamination of groundwater; effluent from farms, septic tanks, or industrial sites is speedily delivered to wells down-gradient.

There is a rich lexicon of karst-related terms from around the world, reflecting countless local variations on the water-limestone pas de deux. From Irish there is "turlough," an ephemeral lake that fills from below when underground rivers flood, then disappears again. The limestones of the Yorkshire Dales are a checkerboard of "clints" (blocky pillars) separated by gaping "grykes" (dissolution-enlarged fractures). German has contributed "karren" to describe the rough, rip-your-pants texture of chemically weathered limestone surfaces. On Mexico's Yucatán Peninsula, there are "cenotes"—caves whose roofs have collapsed, considered sacred to the Maya as sources of fresh water. A semicircular arc of cenotes was a key clue leading to the discovery of the dinosaur-killing impact crater at Chicxulub, Yucatán.

Back in Slovenia, "foibe" refers to deep sinkholes with outward sloping walls—impossible to climb out of—a word forever linked to their use in mass executions in World War II. It seems the lawless nature of a karst landscape inevitably influences the humans who live on it.

See also Darcy; Speleothem; Stygobite

Katabatic Winds
Blow me down

My PhD research focused on the tectonic history of the northernmost extension of the Caledonian-Appalachian mountain belt in Svalbard, Arctic Norway, but fieldwork there taught me much about polar meteorology as well. Most of these weather lessons took the form of immersive tutorials at inconvenient times.

One such session occurred soon after we had set up tents following a long and exhausting hike with full packs to a remote site where we planned to camp and work for a week. We had chosen the spot, in a shallow dip in

the landscape, about a quarter mile from the "snout" of a glacier, because it seemed to be protected from the always bitter winds. We had just shed our boots and were beginning to make preparations for supper, when an eerie roar arose from somewhere up on the glacier. A few seconds later, a gale-force wind hit, so powerful that it was nearly impossible to stand. The wind ripped objects from our hands, rolled our backpacks perilously close to a roaring glacial stream, and pried up one of the tents, even though it was weighted down by sizable rocks. We chased the tent until it tumbled irretrievably over the edge of a steep ravine. By the time we walked dejectedly back to camp, all was still again.

What we had experienced was a katabatic wind, a phenomenon that occurs in arctic and alpine environments when a mass of cold air accumulates, often above a glacier, then comes rushing downslope as a density-driven current. Katabatic winds are to ordinary wind what a laser is to normal light—highly focused and potentially dangerous. The term comes from the Greek for "descending," although that is far too leisurely a description for the violent phenomenon.

In Greek mythology, *katabasis* was the term for a hero's descent into the underworld. After being hit by a katabatic wind, my coworkers and I certainly felt like we'd been to a (cold) version of hell and back. We put our boots and packs back on and moved to a new campsite before the glacier had time to catch its breath for the next laboratory demonstration.

See also Erg; Haboob; Turbidite.

Kimberlite

Diamonds are ephemeral

Among the many minerals forged by the earth, diamonds arguably have the strongest grip on the human psyche. Individual diamonds are like immortal celebrities—they have names (Hope, Koh-i-Noor) and larger-than-life reputations (often malign). Decades of marketing campaigns have brainwashed countless couples into thinking that their relationship is not official without a diamond. But all the mystique humans attribute to diamonds seems frivolous in comparison to their enigmatic geologic origins in a truly bizarre rock called **kimberlite**.

Named for Kimberley, South Africa, the world's leading diamond-producing region, kimberlite is a dense, magnesium-rich igneous rock that shot up to Earth's surface straight from the mantle. No one has ever witnessed the eruption of a kimberlite, but the chemical and physical characteristics of the rock require that it took a high-speed trip from the interior of the earth.

Diamonds are just pure carbon—but with an exceptionally dense crystal structure not formed at low pressures. At surface conditions, graphite—so soft that it rubs off on paper when one writes with a pencil—is the stable crystal form, or "polymorph" of carbon. Making carbon into diamond requires pressures equivalent to depths of 100 miles or more in Earth's mantle. (Large meteorite impacts, which subject rocks to extreme pressures for fractions of a second, can also form diamonds, but these tend to be microscopic—far too small for even a modest engagement ring.) Technically, therefore, diamonds are metastable at Earth's surface—well outside their natural thermodynamic habitat. And contrary to jewelers' assertions that they are "forever," any diamond at atmospheric pressure is converting very slowly to graphite, in atomic-scale layers, from the outside in.

But what would cause diamonds, and the dense kimberlite magmas that bear them, to rise to the surface, passing through lower-density rocks along the way? An important clue is the surprising presence of the mineral calcite in kimberlites. Calcite (calcium carbonate, or $CaCO_3$) is the principal mineral in limestone, precipitated mainly from seawater—commonly by marine organisms—and thereby sequestering atmospheric carbon dioxide in mineral form. The occurrence of calcite in an igneous rock from deep inside the earth indicates that there was abundant carbon dioxide at the site of melting—and this may be the key to transporting otherwise dense magmas up from depth.

Kimberlites occur within pipelike volcanic vents called "diatremes," which are typically a few hundred yards in diameter. They have a fragmental texture, with clasts of varied rock types they apparently picked up on their ascent, all churned together in an unlikely assortment. This suggests that kimberlite magmas were emplaced violently, with turbulent mixing of materials from many depths. Such turbulence points to a pressurized gas phase—presumably CO_2—that impelled the magma to the surface in a natural version of jet propulsion.

But what, exactly, triggers the sudden, localized, gas-powered eruption of a kimberlite remains a mystery. Unlike other volcanoes, which are spatially related to plate boundaries or broad "hot spots" like Yellowstone, diatremes are commonly found in isolation, far from other areas of magmatic or tectonic activity. There are, for instance, Cretaceous kimberlites in Arkansas, a place not generally associated with recent volcanism.

New geochemical analyses of diamonds suggest that in some cases their journeys may be even more extraordinary than previously thought. Although diamonds are made entirely of carbon, not all carbon atoms are alike. Carbon can occur as either of two stable (non-radioactive) isotopes, carbon-12 (^{12}C) and carbon-13 (^{13}C), differing by the weight of one neutron. Most carbon in the mantle—exhaled as CO_2 by ordinary volcanoes—has a known, consistent ratio of ^{13}C to ^{12}C. That value is significantly higher than the $^{13}C:^{12}C$ ratio of carbon that is fixed by photosynthesizing organisms (trees, algae, etc.), who are picky about their carbon and prefer ^{12}C, the lighter variety.

Intriguingly, the $^{13}C:^{12}C$ ratio in some diamonds is very un-mantle-like—that is, so enriched in ^{12}C that it could only have come from atmospheric carbon fixed by photosynthesizing organisms. So somehow—probably via subduction of ocean crust topped by organic-rich sediment—photosynthetically fixed carbon found its way into the mantle, crystallized at high pressures into diamonds, and then one day took a kimberlite pipe (a geologic express elevator) back to the surface.

This triumphant tectonic tale of reincarnation is, however, overshadowed by the tragic human stories linked with diamond mining. Kimberlites are often mined in deep, dangerous open pits, and sales of "blood" diamonds have financed brutal conflict and genocide. Diamonds are a warlord's best friend. Although the international diamond industry adopted the so-called Kimberley Process for certifying the provenance of diamonds in 2003, smuggling, document forging, and inconsistent enforcement have undermined the credibility of the protocol.

How strange that a metastable rock that has no business at Earth's surface has such power to destabilize human affairs.

See also Amethyst; Breccia; Eclogite; Komatiite; Moho.

Klippe [CLIP-uh]
Strata gone astray

From the German for "cliff," a **klippe** is an erosional remnant of a once-continuous slab of rock that was tectonically thrust over other rocks along a gently inclined

fault surface during mountain building (there's also a word for the whole slab: an *allochthon*). Because this type of faulting typically stacks older, deeper rocks on top of younger, shallower ones, a klippe, standing as an "island" of older rocks amid a sea of younger ones, can initially appear to be a violation of the principle that older rocks should be at the bottom of a stratified sequence.

The most famous klippe in North America is probably Chief Mountain, Ninaistaki to the indigenous Blackfeet people, on the eastern edge of Glacier National Park in Montana. A distinctive landmark visible for miles, the blocky peak itself is made of Proterozoic limestone more than a billion years old, while the rock at its base—which provided the slippery surface over which the limestone slid—is Cretaceous shale, a mere 100 million years old.

Erosion can create other interesting geologic geometries in mountain belts where rocks are complexly stacked up on each other by faulting. An erosional hole through an overthrust slab that provides a glimpse of the underlying rocks is called a "fenster" or "window." Any geologist worth her basalt knows the proper answer to the question: "What is the opposite of a window?" A klippe, of course.

See also Allochthon; Geosyncline.

Komatiite [koh-MAT-ee-ite]
The rock that went extinct

According to the principle of uniformitarianism—the idea that the geological "rules" don't change over time—rock types that formed in the past should still be in the process of forming somewhere on Earth today. This is certainly true of most rocks—for example, sandstone, limestone, granite, basalt—but **komatiite** is a notable exception.

Named for the Komati River in South Africa, komatiite is a volcanic rock that only formed on Earth prior to the end of the Archean eon 2.5 billion years ago. The fact that it vanished from the rock record at that time means that the conditions leading to its formation must have ceased to exist. So what changed in the late Archean to shut down global production of komatiite forever? The primary mineral in komatiite—olivine—provides an essential clue.

While some modern basaltic lavas, such as those erupted in Hawai'i, contain olivine, it is typically not an abundant constituent of these young rocks. Moreover, olivine in Hawai'ian basalts is different from the other minerals in that it did not crystallize with them from the molten lava itself but was instead carried up as an unmelted residuum from the site in the mantle where the magma formed. Such crystals are called "xenocrysts" (foreign crystals). In some places, whole chunks of olivine-rich mantle rock, or peridotite, have come up to the surface with the basaltic lava as "xenoliths" (foreign rocks). The famous green sand beaches at Papakōlea on the Big Island are made of olivine xenocrysts and xenoliths that have weathered out of the basaltic lavas that transported them.

Olivine in these Hawaiʻian lavas remained in the solid state because its melting temperature at any given pressure is higher than that of almost any other mineral. Hawaiʻian basalts erupt at about 2,000–2,100°F, which is certainly searing, but not hot enough to liquefy heat-tolerant olivine. The olivine in komatiites, in contrast, bears evidence that it crystallized directly from the molten state. Specifically, komatiitic olivine occurs as long, slender crystals of the sort that form in liquids cooled far below their freezing point. These needlelike crystals—similar to filigreed frost that forms on windows in cold weather—create a texture called *spinifex*, after a grass that grows on the savannas of South Africa. The spinifex crystals in komatiite indicate that the host lavas were erupted at or above 2,900°F, the surface melting temperature of olivine.

Such a high-temperature lava would have had a much lower viscosity than any "colder" lavas of today—comparable to the fluidity of water. Indeed, before they solidified, komatiitic lavas could have flowed turbulently across the land surface for miles, actively eroding their own channels.

The fact that there are no komatiites younger than about 2.5 billion years reflects the long-term cooling of the earth. The young, incandescent Archean Earth had enough heat to melt the mantle completely, producing komatiites; on the older, cooler Earth of today, even at hot spots like Hawai'i, only partial melting occurs, yielding basalt lavas with still-solid lumps of mantle olivine. Earth's internal heat has two origins: primordial heat from the formation of the planet and radiogenic heat produced by ongoing decay of radioactive elements in Earth's interior. Both are waning gradually over time, and at some point in the geologic future, basalt, now the most common rock type on Earth, will also become extinct.

See also Eclogite; Granitization; Kimberlite; Ophiolite; Uniformitarianism; Xenolith.

Lahar [la-HAAR]
Abhorrent torrent

Lahar is a Javanese word for a stealthy and lethal type of volcanic hazard: a dense, fast-moving slurry of water and volcanic ash often described as having the consistency of wet cement. In contrast to the more notorious pyroclastic flows or *nuées ardentes*, whose deadliness is due largely to their searing temperature, lahars are cold and insidious. Lahars can be launched even when a volcano is inactive if heavy rain saturates loose ash on steep slopes, but the most ruinous ones occur during eruptions when a glacier or snowfield suddenly melts. Unlike a jökulhlaup—a glacial meltwater flood that can also be triggered by the dangerous combination of fire and ice—a lahar is equal parts ash and water. Hurtling at up to 30 miles per hour, impelled by momentum from the mass of sediment they carry, lahars highjack ordinary rivers and turn them into unstoppable torrents of sludge.

The deadliest lahar event in history occurred on the night of November 13, 1985, when the Columbian volcano Nevado del Ruiz reawakened, after a century of slumber, in a modest eruption. Plumes of steam and ash were seen emanating from the summit volcano that evening but seemed to pose no immediate threat to surrounding communities far down its slopes. Then at midnight, four massive lahars roared down river valleys on the flanks of the volcano, inundating villages with waves of mud two stories high, and killing more than 25,000 people. The town of Armero, whose rich volcanic soils had been the source of its prosperity, was almost completely buried in lahar-borne volcanic sediment.

In the aftermath of the Nevado del Ruiz catastrophe, geologists around the world developed lahar hazard maps and public education programs for areas around active volcanoes. Thanks to these efforts and to better geophysical monitoring, lahars related to active eruptions are now less likely to cause tragedy since volcanoes tend to send out advance publicity before they launch into their main act.

More sinister, however, are lahars that occur when their host volcanoes are completely quiet. Picturesque towns skirting the base of Mount Rainier in Washington State are awakening to the reality that heavy rainfall or sudden failure of an ash-laden slope could unleash a lahar with no premonitory signals. In fact, many of these communities are built on ancient lahar deposits, a sobering reminder that volcanoes make no distinction between creation and destruction.

See also Jökulhlaup; Lusi; Nuée Ardente; Turbidite.

Lazarus Taxa
Zombie fossils

Charles Darwin dedicated an entire chapter in *On the Origin of Species* to the "Imperfection of the geologic record." His intent was to preempt arguments from skeptics that the fossil record lacked transitional fossils and thus did not support his theory of evolution by natural selection. More than 150 years later, most of these so-called missing links have been found, and geologists have discovered that the rock record is far less "imperfect" or incomplete than Darwin had suggested. Still, there remain some puzzling silences, including those of the "Lazarus taxa"—species or higher-level groups that disappear from the fossil record only to reappear, seeming to rise from the dead, millions or tens of millions of years later.

In some cases, the failure of certain lineages to show up for roll call in the fossil record can be chalked up simply to the low odds of preservation; fossilization is, after all, the exception rather than the rule, especially for soft-bodied organisms with no mineralized shells or skeletons. But when organisms that are normally well represented in the fossil record go missing for long periods of time, their absence may signal something important.

The most dramatic examples of Lazarus taxa are "reef gaps" that follow many of the great mass-extinction events in geologic history: the apparent disappearance of once-abundant reef-building organisms (not only corals but also other calcifying organisms) for millions of years after these episodes of ecosystem collapse. Of these, the longest is the reef gap in the aftermath of the worst mass extinction of them all: the end-Permian event, a near-

death experience for the biosphere. Unlike the more famous end-Cretaceous dinosaur extinction, for which a rogue meteorite can be blamed, the end-Permian cataclysm had an internal source within the earth system. A "perfect storm" of environmental factors—rapid warming, ozone destruction, ocean anoxia, and acidification—seems to have conspired together to devastate both marine and land-based food chains.

In the sea, calcifying organisms, especially corals, were hit particularly hard. Some coral species did go entirely extinct, never to be seen again. But others mysteriously reappear in lower Triassic strata—after being AWOL for *10 million years*. Individual coral animals, or polyps, are tiny organisms that live in symbiotic bliss with even tinier photosynthesizing organisms. In aggregate, these minute teams secrete the mineralized calcium carbonate structures that we think of as coral— and that get preserved in the fossil record. Because they are literally rocky, they are readily and commonly fossilized—except in the presence of acidic water, which can dissolve them.

The absence of coral fossils for millions of years after the Permian extinction suggests that seawater and/or waters in ocean floor sediments were too acidic for coral fossils to form or be preserved—for a very long time. And then one fine geologic day, corals returned. What were they doing during this time? Camping "outdoors" for most of the early Triassic without their mineralized housing? Creating minimalist structures that were too fragile to fossilize? After 10 million years, how were they able to remember how to build a proper reef?

For us in the Anthropocene, the causes of the Permian extinction—including ocean acidification—are uncomfortably familiar, and the story of the post-Permian reef gap reads like a classic good news/bad news joke. The good news is that corals did in fact recover and flourish once again. The bad news is . . . it took a thousand times longer than the entire duration of human history.

See also Anthropocene; Karst; Taphonomy.

Lusi

Its name is Mud

One day in May 2006, outside a small village in East Java, Indonesia, huge volumes of hot mud suddenly began spewing out of the ground onto a rice field near a well being drilled for natural gas. Drillers attempted to quell the eruption by forcing cement, high-density fluids, and even chains of concrete spheres into the well, but the mud kept coming. And it hasn't stopped since. As of this writing, 15 villages have been buried in mud up to 120 feet deep, and 40,000 people have had to relocate. The mud-spewing vent has been named Lusi, from the first letters of *lumpur*, the Javanese word for "mud," and Sidoarjo, the name of the site.

The exact cause for the abrupt appearance of the mud volcano remains controversial. Lusi's proximity to the gas well suggests a connection to the drilling. Moreover, the day that mud began to gush out of the earth, the drilling had reached a depth of almost 9,300 feet and penetrated a limestone layer that was known from other wells to hold overpressured fluids. But muddying that interpretation is the fact that a magnitude 6.3 earthquake had occurred about 100 miles away just two days before Lusi was awakened; earthquakes can pump fluids through faults and fractures in rocks, and some scientists think that this might have been the trigger. Other researchers have sampled the gases emanating from Lusi and concluded that it may in fact be a natural volcanic vent.

The lack of scientific consensus has prevented those whose homes have been lost from claiming compensation from the company that drilled the well. In the meantime, a system of levees and channels has been built to manage Lusi's ongoing outpouring of mud at a rate of about 30 Olympic swimming pools per day. A few enterprising companies have begun to offer guided tours of the vast and desolate mud flat, embracing Lusi as something of an outlaw celebrity whose name is literally Mud.

See also Lahar; Pseudotachylyte.

Moho (Mohorovičić Discontinuity) [moe-hoe-ROE-vi-chick]
Setting boundaries

Portals to the underworld abound in mythology and fiction: Orpheus, in his search for Eurydice, got in through a Peloponnesian cave. On the Big Island of Hawai'i, tradition holds that the deep Waipi'o Valley provides an entry point. And in the imagination of Jules Verne, a volcano in Iceland was a secret doorway to the center of the earth. Although the underworld has always loomed large in the human imagination, it is in reality more inaccessible than outer space. We haven't even made it to the base of the crust—the Mohorovičić discontinuity, or **Moho**.

The radius of the earth is about 3,960 miles (a bit more at the equator, a bit less at the poles—constant rotation gives Earth a slight midriff bulge). In comparison, the deepest mines reach barely 2.5 miles. Even the deepest hole—drilled by the USSR in an absurd subterranean counterpart to the Cold War space race—reached only 7.5 miles, not even halfway through the crust at the drilling site on the Kola Peninsula. At this depth, temperatures were high enough to cause the drill stem

and bit to soften and deform, making it impossible to continue (NATO countries had a parallel drilling campaign in Bavaria but gave up at 6 miles).

Fortunately for geologists, Earth provides some glimpses of what lies beneath. Rocks that were once in Earth's deep crust and upper mantle can be brought to the surface in various ways: erosion may exhume rocks from depth; volcanoes may cough up chunks of unmelted rocks from deep magma chambers; and, on occasion, subduction goes awry and the entire ocean lithosphere—crust and uppermost mantle— are shoved onto the edge of a continent, forming an *ophiolite complex*, where geologists can stroll at leisure across the crust-mantle boundary.

Although such natural exposures are valuable windows into inaccessible depths, most of what we know about the deep interior of the earth comes instead from records of how seismic waves, generated in large earthquakes, travel through rocks. When faults slip in a seismic event, they release energetic waves that ripple through the earth and cause the shaking that we call an earthquake. These vibrations come in several forms, including the fast but relatively benign P- (pressure) waves and the slower but more damaging S- (shear) waves.

Following an earthquake near Zagreb in 1909, Croatian geologist Andrija Mohorovičić noticed that beyond a certain distance from the epicenter, there were two distinct sets of P- and S-wave arrivals on seismic records of the event. He realized this meant that the wave energy took two different paths from the focus of the earthquake: a direct route through rocks of the crust and an indirect route that traveled partly through the upper mantle, which is more rigid and transmits seismic waves faster. Even though the second path was longer, waves taking the mantle route arrived sooner than the direct waves—in the same way that a driver may arrive at a destination earlier by taking an expressway even though the distance is greater than it would be on slow backroads.

Using trigonometry and the time it took the waves to reach different seismic stations, Mohorovičić was able to determine the velocity of seismic waves in the crust (about three miles per second for P-waves), their significantly higher velocity in the mantle (about five miles per second), and the depth to the crust-mantle boundary, which is now named for him: Mohorovičić discontinuity, commonly shortened by geoscientists to the "Moho."

On the continents, the Moho is typically at depths of 20 to 25 miles, although it may be as deep as 45 miles where the crust has been thickened by mountain building. The crust under the oceans is much thinner, and the Moho lies only 3 to 6 miles below the seafloor. Beneath both continents and oceans, the Moho is a profound compositional boundary, separating the relatively light, high-silica rocks of the crust from the much denser iron and magnesium-rich rocks of the mantle.

Although it is a major mineralogical boundary, the Moho is not a significant mechanical one—and, in particular, it doesn't define the thickness of the tectonic plates that shuttle about Earth's surface. Instead, the plates include both the strong crust and the rigid uppermost part of the mantle, which together constitute the lithosphere. In any given region, the base of a lithospheric plate coincides with the depth in the mantle where temperatures approach the melting point of olivine-rich mantle rocks (generally 60–120 miles). Even though only a tiny fraction of the mantle melts at these depths, the rock strength is profoundly reduced. Like the Moho, the base of the lithosphere is easily recognized by a dramatic change in seismic wave velocities—in this case, an abrupt drop—and is thus known, rather prosaically, as the "low-velocity zone."

It is strange to realize that although the Moho and the low-velocity zone lie at distances close enough to be reached in a short drive, or even a bike ride, they will never be visited by humans. And since Orpheus did not take good geophysical notes on his trip, and Jules Verne's account includes dubious mention of giant mushrooms at mantle depths, we must count on seismic waves to be our proxies in exploration of the underworld.

See also Eclogite; Geodynamo; Kimberlite; Ophiolite.

Mylonite [MY-luh-nite]
Faulty logic

Although it was coined by geologists in the late 19th century, the term ***mylonite*** shares an ancient Indo-European root with "mill, meal, molar, maelstrom," and even "Mjölnir" (Thor's hammer), meaning "to grind or crush." Mylonites are fine-grained rocks found in fault zones, so this etymologic allusion may seem apt, but it was actually based on early misconceptions about how such rocks form.

One typically thinks of faults as brittle discontinuities in rock. But a broader, more accurate definition of a fault would be a zone along which significant displacement has occurred. This could happen by brittle failure, fragmentation, and loss of cohesion, or by ductile flow, which is perhaps akin to how the mascarpone layers in a tilted tiramisu might slide past one another. Mylonites form in the latter manner—through ductile fault displacement—but at temperatures far warmer than tiramisu should ever be served.

In the upper part of continental crust, where most earthquakes occur, rocks are brittle and actually grow stronger with depth owing to progressively higher pressure, which increases the frictional resistance across fractures and other potential failure surfaces. Temperature also increases with depth, at a typical geothermal gradient of about 75°F/mile. Temperature does

not much influence the strength of rocks—until they find themselves at depths of about 10 miles, where it becomes possible for minerals like quartz and feldspar to flow plastically as solids through molecular-scale deformation. Below this "brittle-ductile" transition, rock strength falls abruptly, earthquakes are rare, and weak surfaces in rocks accommodate tectonic strain by shearing and recrystallizing, forming mylonites.

Early geologists assumed that the fine grain size in mylonites was a product of physical crushing, but thanks to advances in material science, we now understand that mylonites form in a natural analog to the metallurgical process of "hot working," in which dramatic shape changes can occur without fracture if temperatures are high enough for continuous solid-state recrystallization to occur. The ongoing formation of tiny, new, strain-free crystals as shearing progresses is what creates the fine texture of mylonites.

Through such crystal plastic "hot working," deep crustal mylonite zones along plate boundaries are able to keep up with the stately ambient pace of tectonics, measured in inches per year. Meanwhile, in the shallow, brittle crust, friction impedes slip, causing faults to get stuck and steadily accumulate stress—until at last they can't take it anymore. Then they fail suddenly in earthquakes, lurching at rates of three to five feet per *second*. Paradoxically, it takes strong rocks to fail in earthquakes; weak ones simply ooze, never building up enough stress to be released in sudden seismic failure.

There are sites around the world—coastal Maine, northern Wisconsin, the Outer Hebrides—where, thanks to erosional exhumation, it is possible to study ancient fault zones that were once at the depths of the critical transition from brittle and seismic to plastic, nonseismic behavior. Intriguingly, at a few of these places, rocks diagnostic of ancient earthquakes, called *pseudotachylyte*, formed by frictional melting, are found to both cut across, and also be cut by, mylonites. These mutually crosscutting relationships indicate that when the ancient faults were active, they alternated between slow-motion plastic deformation and episodic, lightning-quick seismic slip—probably when large earthquake ruptures that began in the upper crust barreled down into the depths where ductile behavior normally occurs.

There's a groanworthy joke passed from one generation of geology students to the next that goes: "How fast does a fault slip?" Answer: "About a mile a night."

In fact, that's a little slow, at least for those seismically jolted mylonites. A mile in an eight-hour night translates to about 0.2 feet/second (even slower on longer winter nights)—not quite up to earthquake speed.

See also Benioff-Wadati Zone; Breccia; Pseudotachylyte; Slickensides.

Namakier [NAM-a-keer]

Salt − water = taffy

From Farsi for "mountain of salt," a *namakier* is perhaps more accurately described as a glacier of salt that can flow over the land surface as fast as inches per year, rivaling its icy counterparts for speed. Although "salt of the earth" is often used to describe ordinary people, the actual salt of the earth is anything but an ordinary rock.

Rock salt forms in lagoons or other isolated basins when briny water evaporates to the point where it can no longer hold its dissolved solids. Minerals, including halite (sodium chloride or table salt), sylvite (potassium chloride, used as a salt substitute), and gypsum (hydrous calcium sulfate, aka plaster) precipitate out in a crystalline mass like (very salty) rock candy.

When first deposited, these "evaporite" minerals are denser than sediments such as mud or sand that might be deposited on top of the salt. But as the overlying strata themselves become buried, their interstitial pore spaces collapse and their density steadily increases. Crystalline salt, in contrast, lacking open voids, resists compaction. So at some point, typically about a mile below Earth's surface, salt becomes more buoyant than the rocks above it, and then its other extraordinary tendency—its capacity to flow viscously in a solid state—is awakened.

The buried salt becomes restless, seeking to correct the density inversion, like a blob of wax rising in a lava lamp. It pushes surrounding rocks out of the way, creating a strange subsurface architecture of salt domes, pillars, and canopies. Mobile masses of salt create traps for petroleum and are the primary targets for oil exploration in the Gulf of Mexico and elsewhere.

Under special geologic conditions, an eroded salt dome may spill out onto the land surface as a namakier. Just as ice glaciers can exist only in regions cold enough for snow to persist, salt glaciers can exist only where it is *dry* enough for salt to resist being dissolved away by rain. The Zagros range in Iran is one of the only places on Earth where ancient geology and modern geography converge to create namakiers—the mountains that flow.

See also Deborah Number; Lusi; Pingo.

Nuée Ardente [NOO-ay ar-DAHNT]

Cloud with a white-hot lining

From the French for "burning cloud," a **nuée ardente** or pyroclastic flow, is the most dangerous of all volcanic phenomena. A roiling mix of hot gases and incandescent ash, a nuée ardente can career down the slope of a volcano as fast as 70 miles per hour, often entraining large boulders at its base. In the AD 79 eruption of Mount Vesuvius, the cities of Pompeii and Herculaneum were overrun by nuées ardentes, and recent archaeological work suggests that the primary cause of death for many victims was not asphyxiation as once thought but, horrifyingly, boiling of the blood due to intense heat.

The first modern scientific description of the terrible power of a nuée ardente was made by French geologist Alfred Lacroix based on accounts by shipboard witnesses of the ruinous 1902 eruption of Mount Pelée on the island of Martinique in the West Indies. The volcano is part of the Lesser Antilles island arc, a subduction-related chain of volcanoes, where the South American tectonic plate is sliding under the Caribbean plate. Its 1902 eruption remains one of the deadliest in recorded history.

In the weeks preceding the disaster, Pelée had been showing signs of reawakening. There were frequent steam eruptions and earthquakes, many of which triggered avalanches. Birds fell from the sky, apparently choked by ash or poisoned by volcanic gases. Snakes and giant centipedes, shaken out of the forests around the volcano, menaced people in the city of Saint-Pierre. On May 5, 1902, a thick volcanic mud flow or lahar swept down from the summit of the volcano, hijacking the channel of a major river and killing 20 people at a sugar mill near the coast. Three days after that, Pelée unleashed the infamous nuée ardente that completely destroyed Saint-Pierre, killing virtually all of its 30,000 residents in three minutes.

One of the few survivors was a prisoner in an underground cell, a man named Louis-Auguste Cyparis (also recorded as Ludger Sylbaris). Cyparis was rescued four days after the catastrophe, and in light of the trauma he had suffered, granted a pardon and released. As the world learned of the tragedy in Saint-Pierre, Cyparis became an international celebrity and toured with the Barnum and Bailey Circus, appearing before audiences in a replica of his prison cell.

Indirectly, the eruption of Mount Pelée continues to influence global shipping today. At the time of the eruption, the best location for a canal through Central America was still being debated. The violence of Pelée's eruption made siting the canal in volcanically active Nicaragua seem unwise, and Panama was chosen instead.

As devastating as the Mount Pelée nuée ardente was, it has arguably saved lives in subsequent decades because the event marked the beginning of modern systematic volcanology. Alfred Lacroix's detailed reconstruction of the timeline of the scalding torrent laid the foundation for studies that helped to inform evacuation plans prior to the eruptions of Mount Saint Helens in 1980, the Philippines' Mount Pinatubo in 1991, and other more recent volcanic events.

In addition, the deposits left by the Mount Pelée nuée ardente have served as a kind of Rosetta stone, which has made it possible for geologists to decrypt the rock record of ancient volcanic eruptions and document processes that cannot safely be observed in real time.

This has led to the recognition of volcanic eruptions in the geologic past that are orders of magnitude larger than any witnessed in human history. For example, the rocks at Yucca Mountain, Nevada, once the proposed site for the US high-level nuclear waste repository, were formed by a series of gigantic pyroclastic flows some 15 million years ago, far bigger than that at Mount Pelée or any modern volcano. Although the repository project is now indefinitely tabled, there is some metaphorical logic to using an ancient "burning cloud" to protect us from hot radioactive waste.

See also Benioff-Wadati Zone; Lahar; Rapakivi.

Nunatak [NOON-ah-tuck]
Peaks in a blanket

An Inuktitut and Greenlandic term, **nunatak** describes a rocky peak peeking out from beneath a cover of glacial ice. In arctic terrains with few other landmarks, the distinctive shapes of particular nunataks have long been vital navigational guides.

Nunataks may have been the inspiration for the humanoid way-finders called *inuksuk* that Arctic people have built from stacked stones for centuries. An inuksuk is featured on the flag of the Arctic Canadian province of Nunavut, whose name shares with nunatak the root *nuna*, meaning "land." Nuna is also the name given, posthumously, to a supercontinent that was assembled about 1.8 billion years ago, with the Canadian Shield region as its heartland (a great-grandparent of the pipsqueak Pangaea, formed just 280 million years ago).

See also Firn; Pingo; Polynya.

Nutation

Nodding off

Most Earthlings—and any extraterrestrial observers who may be watching this planet—are familiar with the basic choreography of its motions through space: its stately daily rotation and reliable annual circuit around the sun. But Earth is, in fact, a bit of a whirling dervish, or perhaps a bobble-head, rocking, bowing, and dipping at different rates even as it revolves and orbits.

Nutation, from the Latin word for "nodding," is one category of such motions—slight changes in the orientation of the planet's rotation axis caused by the gravitational effects of the changing positions of the sun and moon acting on a planet with a nonuniform distribution of mass. Nutations are much smaller and quicker than the 10- to 100,000-year changes in Earth's tilt and orbit known as the Milankovitch cycles, which are tai chi–like in comparison (and of long enough duration to incite ice ages).

Earth's primary nutation cycle elapses over 6,798 days, or almost 19 years, but superimposed on that are smaller and higher frequency dips that occur over months or days. These shorter cycles reflect the shifting mass of the atmosphere and oceans as air and water are churned by seasonal weather patterns, Coriolis vortices, and world-circling currents. One of Earth's best-known dance moves, the 14-month "Chandler Wobble," is attributed to regular oscillations in pressure on the deep seabed. Occasionally, very large earthquakes change Earth's tilt by a tiny fraction of a degree in a single jerk.

As implausible as it may seem, we humans have, unintentionally, altered Earth's motions in space. The massive Three Gorges Dam across the Yangtze River has impounded so much water at a higher-than-normal elevation that it has slowed Earth's rotation rate by a small amount. And as anthropogenic climate change continues, polar ice sheets will lose increasing amounts of mass to the world's oceans, altering Earth's nutational cadences. Even from a great distance, an astute alien onlooker would be able to infer that something had caused the planet to lose its old groove.

See also Geodynamo; Sverdrup.

Oklo [AWK-low] Natural Nuclear Reactor
The original Manhattan project

Chernobyl and Fukushima are names one might first think of when the cheerful subject of runaway nuclear fission reactions comes up in conversation. But

two billion years before either of these tragic accidents, a natural nuclear "meltdown" occurred at a place called Oklo in what is now Gabon in West Africa. This strange event could only have happened at one particular moment in Earth's history, when two unrelated geochemical trends converged in a critical way. While the Oklo "reactor" is apparently a unique phenomenon, the story of how it happened sheds light on the long-term evolution of the earth as a whole.

The first geochemical trend that led to the chain reaction at Oklo had to do with the gradually changing ratio of the two main isotopes of uranium, ^{235}U and ^{238}U. Both isotopes are radioactive, transmuting over time (measured in half-lives) into other elements by emitting radiation. However, only ^{235}U is fissionable—that is, capable of spontaneously splitting in half and thereby releasing large amounts of energy. At present, ^{238}U accounts for 99.3 percent and ^{235}U for only 0.7 percent of all uranium on Earth, owing to the different primordial amounts of the two isotopes and their distinct half-lives (^{235}U was initially less abundant and also decays more quickly than ^{238}U). Moon rocks and meteorites have the same proportions of the two uranium isotopes since these bodies formed from the same starting ingredients as Earth in the early solar system.

Using uranium for nuclear power (or bombs) requires "enriching" it so that it has at least a few percent of the fissionable ^{235}U. Until about 1.5 billion years ago, before much of Earth's original inventory of ^{235}U had decayed away, the ratio of ^{235}U to ^{238}U would have been high enough for a fission chain reaction to begin—*if* enough uranium had been concentrated in one place, in what Manhattan Project scientists dubbed a "critical mass." And this is where the second geochemical plotline enters the Oklo story.

Uranium is common in crustal rocks, especially granites (yes, your countertop is slightly radioactive) in concentrations measured in parts per million—not nearly enough to reach a critical mass, nor even to constitute a minable uranium ore body. But nature has a mechanism for gathering these rare atoms together in larger numbers: water moving through rocks can collect uranium atoms in solution, like a bus steadily filling up with passengers. Groundwater can only pick up uranium "passengers," however, if it contains dissolved oxygen; otherwise uranium is insoluble and stays in its host rock.

Before the Great Oxygenation Event 2.45 billion years ago—a first-order geochemical inflection point in Earth's history when photosynthetic microorganisms introduced free oxygen (O_2) into the atmosphere—surface and groundwater had too little oxygen to dissolve uranium from rocks. But once oxygen became available at Earth's surface, the geochemical rules changed. Groundwater trickling through granitic rocks could acquire significant amounts of uranium and carry it away, at least up to the point where the water encountered lower oxygen conditions, at which point the uranium would be abruptly precipitated in high concentrations—like a busload of passengers forced to disembark at the end of the line.

The natural fission reactor at Oklo could only have happened in the brief geologic window of time when: (1) the ratio of ^{235}U to ^{238}U had not fallen below the "enriched" value and (2) the amount of oxygen in the atmosphere had risen enough for groundwater to be able to create "critical mass" concentrations of the element. The process would have started when some of the ^{235}U in the water-deposited ore body spontaneously split, emitting high-energy neutrons that struck other ^{235}U atoms, causing them to fission and so on. Groundwater in the rocks apparently acted like the "moderator" medium in a nuclear power plant (usually graphite or "heavy water"), preventing the reaction from causing a massive explosion.

The events at Oklo can be reconstructed in detail by analyzing the modern concentrations there of uranium isotopes and those produced by fission of ^{235}U. In fact, the natural reactor was first recognized in 1972 by a French nuclear physicist who noticed that the ratio of ^{235}U to ^{238}U in uranium ore mined at Oklo was lower than that in any other rock ever analyzed—including moon rocks and meteorites. He realized that the only explanation for this was that a significant amount of ^{235}U has been "lost" in a natural chain reaction.

Subsequent work at Oklo suggests that the process started and stopped multiple times at different spots within the ore body and continued over a period of hundreds of thousands of years. Whether the nuclear reactions had any effect on living organisms at the time is unknown; the biosphere of the early Proterozoic was entirely microbial. Certainly the site would have been uncomfortably hot, and radiation from the many short-lived isotopes produced in the fission reactions could have damaged cellular DNA.

Although evidence for similar nuclear chain reactions has not been found anywhere else, it is possible that there were such "natural reactors" in other parts of the world during this singular interval in geologic time, and that erosion, burial, or subduction have destroyed the record of them. More than a mere geological oddity, Oklo is a reminder of Earth's long and labyrinthine biography, with many complex story lines—geochemical, hydrological, biological, and tectonic—unfolding and intersecting over geologic time.

See also Acasta Gneiss; Geosyncline; Zircon; Simplified Geologic Timescale (Appendix 1).

Ophiolite [OH-fee-oh-lite]
Crocodile rock

Literally "serpent stone," an ***ophiolite*** is not a single rock type but a distinctive association of rocks commonly found together in the interior of mountain belts. Although ophiolites were described by geologists in the early 19th century, their origin was an enigma until the late 1960s. The name was coined around 1820 by French polymath Alexandre Brongniart, whose primary scientific work was the study of reptiles. Perhaps it's not surprising, then, that he saw reptiles everywhere, even

in snow-covered rocks high in the Alps. But Brongni-art was not alone in sensing something herpetologi-cal about ophiolites; the Lizard Peninsula in Cornwall hosts a world-famous example.

Ophiolites include bodies of olivine-rich peridotite, which is the most abundant rock on Earth because it makes up much of the mantle. But peridotite is far outside its thermodynamic comfort zone when it lies at or near Earth's surface, and it will react readily with water and carbon dioxide to form new minerals. Peridotite outcrops are almost always partly altered, through interaction with water, to a shiny, slippery, dark green mineral evocatively called "serpentine"—and this was probably Brongniart's inspiration for the name "ophiolite." But another rock type characteristic of ophiolites, pillow basalts—formed when lavas are extruded underwater and cool into bulbous shapes—can also resemble a writhing pile of snakes, so Brongniart's neologism is doubly apt.

A third rock type found in most ophiolite complexes is chert (or chalcedony)—a finely crystalline form of SiO_2, with the same chemical composition as the mineral quartz. Unlike most quartz, which can ultimately trace its origin back to an igneous rock, chert is a sedimentary mineral most commonly formed from the shells of minute eukaryotic organisms, especially radiolarians and diatoms. Far from continents, almost no land-derived sediment accumulates on the seafloor; instead, the fine rain of these tiny mineralizers onto the seabed is the main type of deep marine deposits, forming a quivery mass called "pelagic ooze," which eventually solidifies into chert.

The fact that this odd triumvirate of rock types—mantle peridotite, submarine basalt, and microfossil-rich chert—occurred together in mountain ranges around the world caught the attention of German geologist Gustav Steinmann (1856–1929). (Incidentally, Steinmann's surname, meaning "stone man," is a fine example of nominative determinism.) The unlikely rock trio became known as "Steinmann's Trinity," but it would be decades before anyone could explain *why* the three rock varieties would tend to hang around together and how they could end up at the summits of mountains.

In 1968, British geologist Ian Gass, who had spent years studying an ophiolite complex on Cyprus known

as the Troodos Massif, argued that it represented ancient oceanic lithosphere—the deep seafloor, all of the
crust and even some of the upper mantle—that had
been thrust up onto land where geologists could marvel
at it, rather like a giant deep-sea creature rarely glimpsed
in life. In fact, Gass's detailed observations of features in
the Troodos Ophiolite caused many previously skeptical
contemporary geologists to accept the then hypothetical
process of seafloor spreading, which had been proposed
a few years earlier based on indirect observations but
not yet observed firsthand (see *Eclogite* for a description
of the phenomenon).

Ophiolites in mountain belts elsewhere in the world—
Switzerland, Oman, Newfoundland, New Zealand—
were soon understood to represent the only remaining
traces of vanished oceans that had otherwise been swallowed by subduction in the run-up to continental colli-
sion. Geologists today are still arguing over the reasons
why some ocean lithosphere resists its destiny and re-
fuses to be subducted. Maybe Brongniart had his herps
wrong; ophiolites are actually more like amphibians
than reptiles—born underwater but later emerging onto
the land.

With their high levels of magnesium, nickel, and
chromium, peridotites create soils that are toxic to most
plants, so areas where ophiolites occur are often barren, otherworldly looking landscapes. But their unusual
chemistry could provide a partial antidote to rising atmospheric carbon dioxide levels: these mantle rocks, so far
from their native habitat, are quick to assimilate any CO_2
dissolved in rain and groundwater, and they lock it away
in mineral form as magnesite (magnesium carbonate

or $MgCO_3$). Some geoscientists are investigating this process as a long-term carbon sequestration strategy. The main, and not inconsiderable, difficulty with this scheme is getting concentrated CO_2 to the far-flung sites where ophiolites occur so we can feed it to the reactive "snake stones." For the time being, however, it would be much better to curb our own carbon appetites.

See also Benioff-Wadati Zone; Eclogite; Kimberlite; Komatiite; Moho.

Panthalassa [pan-thuh-LASS-uh]
Oceans of time

As land-dwelling creatures, we tend to view the world as land surrounded by sea, when in fact it is sea sprinkled with land. So while you're probably familiar with Pangaea ("all Earth"), the supercontinent that formed in the late Paleozoic era and then broke up during the Mesozoic, it's likely you've never heard much about the other side of the world at the time Pangaea existed: that was ***Panthalassa***—"all ocean."

Apart from our intrinsic terrestrial bias, reconstructing the paleogeography of the earth in the geologic past is inherently land-centered because of a fundamental distinction between the crustal material that underlies continents and oceans. Continental crust, made, broadly speaking, of granite (but in detail very heterogeneous) can persist for billions of years. All of the modern continents have central stable areas called "shields" that formed in Archean time more than 2.5 billion years ago. Ocean crust, in contrast, rarely spends more than about 180 million years at Earth's surface. Born as incandescent lava at submarine volcanic fissures called "mid-ocean ridges," ocean basalt does little during its life span other than quietly cool and contract. Virtually all ocean crust has a uniform and predictable destiny dictated by its steadily increasing density—subduction, or slow reassimilation into the mantle from which it came. (The only exceptions are ophiolites—slabs of ocean crust that rebelled against this predetermined fate.)

As a result of this asymmetry in the life expectancy of the two types of crust, drafting paleogeographic maps for times before about 200 million years ago is like putting together a jigsaw puzzle from which two-thirds of the pieces are missing. Ancient continental positions can be recreated using a combination of fossils diagnostic of certain climates, the similarity of rock sequences in different parts of the modern globe, the continuity of ancient mountain belts, and paleomagnetic signatures in iron-rich rocks, which provide information about the latitude at which they formed.

In contrast, since there is almost nothing left of the contemporary oceans of the time, their presence can only be inferred. Still, we know they were there and even give them posthumous names. Iapetus, which existed between about 600 and 270 million years ago, is one such ghost ocean, though it didn't span an entire hemisphere, as Panthalassa did. The closure of Iapetus (through subduction) formed Pangaea, and as the precursor to the modern Atlantic, it is named for the father of Atlas, a Titan in Greek mythology. The Mediterranean Sea is the last vestige of a once-vast ocean called Tethys, for the Titan goddess of water. The Alps, Carpathians, and Himalaya mark the sites where Tethyan ocean crust has been consumed by subduction.

From formation to final breakup, supercontinents have life cycles of 500–700 million years; Pangaea was preceded by Rodinia (at its peak about 1 billion years ago) and the even more ancient Nuna (1.8 billion years ago). Each of these would have had its complementary "superocean," meaning that Panthalassa, too, is just the most recent in a long line of titanic ancestral seas, stirred by forgotten currents, wracked by unrecorded hurricanes, teeming habitats for long-departed ecosystems.

As writer Arthur C. Clarke observed, "How inappropriate to call this planet Earth when quite clearly it is Ocean."

See also Acasta Gneiss; Benioff-Wadati Zone; Granitization; Ophiolite; Simplified Geologic Timescale (Appendix 1).

Pedogenesis [ped-uh-GEN-eh-siss]
Dirt rich

Pedogenesis might sound like the spontaneous sprouting of feet—or possibly of children (in the manner of some Greek gods). But pedogenesis is in fact the far more gradual process of the formation of soil—one of the most sophisticated, and least appreciated, of Earth's many innovations.

The proposal that we might be able to "terraform" another planet invariably captures the public imagination; depictions of a green Mars are an enticing mashup of myths ranging from the Garden of Eden to the American frontier and *Star Trek*. And what a good idea to have a planet in reserve in case we really mess this one up! One problem, however, is that even if we could homestead on a new planet, we would still be us—the same

flawed creatures expelled from Eden. Another little challenge would be the lack of soil—also known, tellingly, as earth—that would allow us to grow a new garden.

Proponents of extraterrestrial colonies fail to appreciate that so many of the planetary amenities we take for granted at home on Earth are simply not going to be available on planets with very different histories and habits. Soil, in particular, is a very earthy combination of living, once-living, and once-rocky matter, all in various stages of decay and disintegration. In the days of the Apollo missions, NASA sometimes used the word "soil" to describe the powdery veneer of pulverized rock on the moon's surface, the product of four billion years of meteorite bombardment. (The unknown physical properties of this "rock flour" were a major concern before the first moon landing: some NASA engineers worried that the lunar module might sink irretrievably into it.) But to call this sterile, shock-blasted material "soil" is perhaps akin to smashing up some ice cubes and telling guests that you are serving gelato.

The primary mineral components of true soils—clays—are formed over time when igneous rocks, particularly granite (unique to Earth), are weathered by abundant liquid water (unique to Earth) in collaboration with various organic acids (you might be noticing a pattern). Exactly how long it takes to break raw igneous minerals down into tillable soil depends on climate, erosion rates, and many other factors, but millennia would be a conservative estimate.

Clays are a large family of minerals made primarily of aluminum, silicon, oxygen, and hydrogen—the residuum left when more soluble elements like calcium and potassium have been coaxed out of rocks after long exposure and carried away by rainwater. Clay minerals typically occur as tiny crystals that have large surface areas bristling with unbonded atoms, and this causes them to be especially reactive.

At the molecular scale, all clays are organized into silicon-rich ("bread") and aluminum rich ("peanut butter") layers, and the various species of clay minerals differ mainly in the way that these layers are organized. For example kaolin, or the white "china clay" used in porcelain dishes and ceramic tiles, has a crystal structure like a stack of open-face sandwiches, while most other clays, including vermiculite, commonly used in potting soil, are like piles of closed-face sandwiches. These closed-face types differ further according to whether the "bread" slices are touching each other or (and the sandwich metaphor gets soggy at this point) can be lofted apart by layers of water molecules. The clay minerals with the greatest capacity to take in water are smectites, or swelling clays, whose tendency to expand and contract can wreak havoc with roadbeds and building foundations. The particular mix of clay species, sand or rock fragments, and organic matter—all of which are legacies of the region's geological, biological, and hydrological history—give any given soil its characteristic texture or "tilth."

Although no two soils are exactly alike, pedologists, or soil scientists, have developed elaborate classification schemes to impose some sort of order on the profusion of types. The US Department of Agriculture has a hierarchical Linnaean-like system in which soils are classed into order, suborder, great group, subgroup, family, and series. There are 12 orders, including the mellifluous "mollisols" (soft, dark, organic-rich soils typical of undisturbed grasslands) and the immature "inceptisols" (not too far removed from the "parent" rock). At each subordinate level, these names are further embellished with adjectives and prefixes, leading to Esperanto-like neologisms like "thaptohistic cryaquollic mollisol." One wouldn't necessarily expect people who muck about in the dirt to be punctilious linguists, but the USDA soil taxonomy website provides a helpful pronunciation guide with etymological notes.

While most pedologists focus on modern soils, ancient soils are the subject of the relatively new subdiscipline of paleopedology (a sesquipedalian word meaning neither the study of old children nor of old feet). Paleopedologists have found that soils as we know them are relatively recent arrivals on the geologic scene; before the first plants and fungi moved onto land in the Ordovician period, about 450 million years ago, soils were thin, stony, and easily washed away by rivers that were likely less channelized in the absence of vegetation. And during most of Precambrian time, when the atmosphere had a Mars-like carbon dioxide–dominated composition, soil chemistry was acidic and alien. Earth itself wasn't "terraformed" then.

Earth's recipe for dirt is billions of years in the making and won't be easily reproduced anywhere else. We should never forget that soil is a miraculous, mysterious, semiliving, sacred substance—even when tracked into the house on children's feet.

See also Amethyst; Areology; Geodynamo; Grus; Oklo.

Pingo

Young as the hills

Like nunatak, ***pingo*** is an Inuktitut term for a distinctive polar landform—in this case, an ephemeral ice-cored hill. To me, the name seems onomatopoeic, evoking a pingo's curious ability to spring up and then vanish again.

Pingos most commonly form in polar wetlands when water freezes, expands, and causes the overlying land to dome up. Such pingos can wax and wane seasonally—like giant goose bumps that form on the land in the chill of winter and disappear in the warmth of summer. Another,

slower mechanism by which pingos form involves over-pressured groundwater that rises and freezes, creating an ice barrier that leads to even higher water pressures, in a positive feedback process—a (subterranean) snowball effect. This type of pingo can grow several inches per year and persist for centuries. Sadly, however, as global climate warms, all pingo "species" may soon be extinct.

See also Jökulhlaup; Lusi; Nunatak; Polynya.

Pneumonoultramicroscopicsilicovolcaniconiosis [pronounced exactly as spelled!]

Ashes, ashes

A performance piece as well as a technical term, ***pneumonoultramicroscopicsilicovolcaniconiosis*** is a deadly lung condition resulting from inhalation of volcanic ash—essentially tiny shards of glass.

Magma is typically a three-phase mixture, consisting not only of molten rock but also some crystals and dissolved gases. As magma rises and nears Earth's surface to become lava, gas bubbles exsolve from it just as they do when a champagne bottle is opened. Initially this forms a hot foam, which may survive as pumice—but if the bubbles expand very rapidly, the entire mass of magma is blown apart, creating ash. Under the microscope, individual ash particles commonly have the form of tiny triangles with concave sides. This distinctive, spiky shape reflects their origin at the interstices between clusters of bubbles—and makes ash particularly damaging to the lungs.

Once one masters its pronunciation, one naturally seeks opportunities to slip pneumonoultramicroscopicsilicovolcanoconiosis into everyday conversation. For

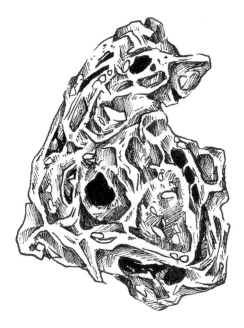

example, "Say, did you hear what killed all those Miocene rhinoceroses at Ashfall Fossil Beds State Historical Park in Nebraska?" As a way to entertain ourselves one summer in a remote camp on Ellesmere Island in Arctic Canada, my geologic colleagues and I tried to come up with the most confusing possible radio alphabet—an utterly impractical replacement for the standard Alpha-Bravo-Charlie-Delta-Echo, and so on. The words for the letters G, K, and N were Gnu, Knew, and New, and for P, of course, pneumonoultramicroscopicsilicovolcanoconiosis.

See also Amygdule; Nuée Ardente.

Polynya [poh-LYN-ya]
Window of opportunity

Of Russian origin, **polynya** means "a rare area of open water surrounded by sea ice." Formed by particular combinations of wind and upwelling currents, polynyas tend to persist for decades and were traditionally important food sources for people of the Arctic. Like oases in the desert, such sites are biodiversity hot spots in regions that are otherwise barren of life. Plankton blooms and shoals of krill attract fish, which in turn bring birds, seals, walruses, and polar bears.

During the deep ice age called Snowball Earth or the Cryogenian period 720–635 million years ago, when there were glaciers at the equator and the world's oceans were covered with ice, polynyas may have been the only places where the habit of photosynthesis was maintained—rather like monasteries in the Dark Ages, where the flame of literacy was kept alive.

Today, as arctic sea ice dwindles owing to global warming, polynyas are both growing and, as a result, paradoxically, disappearing. Polynyas cannot exist without their polar opposite—the enclosing ice.

See also Cryogenian; Nunatak; Pingo.

Porphyry [PORE-fer-ee]
All crystals great and small

A **porphyry** is an igneous rock with large crystals called *phenocrysts*—typically of blocky feldspar—set in a fine matrix of other minerals. The word comes from the Greek for purple, even though most porphyries are not. Purple was the rich hue, however, of Imperial Porphyry, mined (mainly by slaves) in Egypt for the Roman elites

and used for panels in the Pantheon, as well as pillars, statues, and monuments around the empire. Centuries after the fall of Rome, the name persists, although the term "porphyry" is now used to describe a rock texture rather than a color.

For most igneous rocks, crystal size is inversely related to the rate at which they cooled from a magma. If molten rock cools slowly underground, insulated from cold surface conditions, there is time for large crystals to grow. Depending on the magma's composition, this will produce coarse-grained rocks like granite or gabbro, whose crystals can be seen with the unaided eye. But if magma is erupted at the surface, thereby becoming lava, it will cool far more quickly, forming finely crystalline rocks like rhyolite or basalt—or in the case of very sudden quenching, noncrystalline or glassy rock: obsidian.

According to this logic, with their two distinct grain sizes, porphyries would seem to be self-contradictory—the big phenocrysts asserting, "We grew large at a leisurely pace," while the fine-grained minerals testify, "We barely had time to take shape!" In fact, both groups are telling the truth, and this reflects a subtler fact about how magmas solidify. Different minerals "freeze" out of a magma at different temperatures—a phenomenon called "fractional crystallization." Minerals with the highest crystallization temperatures (future phenocrysts) will nucleate first, invariably getting a head start on those that form only at lower temperatures (the undifferentiated matrix).

In some strange way, the Imperial Porphyry can be read as an igneous metaphor for class disparities in Roman times. And just as we still use Roman nomenclature, we can't seem to shake some of the worst attributes of Roman culture.

See also Granitization; Pseudotachylyte.

Pseudotachylyte [soo-doh-TACK-uh-lite]
Hot flashes
An unnecessarily opaque term that causes even seasoned geologists to furrow their brows in incomprehension, *pseudotachylyte* is the word used for veins of dark, glassy rock produced by frictional melting along a fault during an earthquake. The prefix "pseudo," of course, means "false" or not genuine, but hardly anyone knows what on Earth the word "tachylyte" means. As it happens, tachy-

lyte is a rarely used term for basaltic obsidian (volcanic glass). Pseudotachylyte, in contrast, is found in rocks that are clearly not volcanic—and weren't in fact anywhere near Earth's surface at the time they formed.

Although its name is rather obtuse, pseudotachylyte itself is an important source of information about what actually happens at the origin, or hypocenter, of an earthquake. Most earthquakes begin when a segment of a fault begins to slip, usually at depths of at least three miles in the subsurface. Modern seismic networks instantly capture and analyze the resulting waves that emanate from the site of rupture, but there is no way to directly observe the physical processes that occur along a fault during an earthquake.

There are many other types of rock found in fault zones—for example, breccia and mylonite—but these are not necessarily records of slip in an earthquake; they could be formed by slow, nonseismic creep on faults at rates of inches per year. The frictional melting that forms pseudotachylyte, however, requires a dramatic, localized spike in temperature, and this can only happen at seismic slip rates—several feet per second or faster. Pseudotachylytes are thus "fossilized earthquakes," providing rare eyewitness accounts of subterranean processes inaccessible to surface-dwelling humans.

See also Breccia; Mylonite; Slickensides; Twist Hackle.

Rapakivi [RAH-puh-kee-vee]
Oddballs

Rapakivi is a euphonious Finnish word with a malodorous meaning: "rotten rock." Geologists use the term to describe the distinctive texture of a type of

granite that occurs most famously in Southern Finland and is marketed around the world (usually as "Baltic Brown") for stone facades and countertops. You've almost certainly seen this rock; it's as striking and stylish as anything from the Finnish design-house Marimekko—contrary to its less than glamorous Finnish name.

Most granites have blocky, interlocking crystals of pink potassium feldspar (orthoclase) and white sodium feldspar (albite), interspersed with black hornblende and biotite mica. Rapakivi granites have the same minerals, but in a remarkable configuration: the feldspars are almost perfectly round, with large pink cores of orthoclase rimmed by narrow white "halos" of albite. Flecks of black hornblende commonly occur in concentric bands within the pink interiors. On cut and polished surfaces, the rock looks like a collection of oversized marbles or jawbreakers seen in cross section. The only thing rotten about rapakivi is that when it's been exposed to enough

Baltic winters, it tends to disintegrate into a rubble of these crystalline spheroids.

The origin of rapakivi's eye-catching pattern is not fully understood, but it almost certainly reflects some sort of magmatic disequilibrium. Governed by the logic of their underlying molecular structure, feldspars normally form angular, flat-sided tabular crystals (indeed, *orthoclase* means "straight breaking"). The nearly spherical shape of the feldspars in rapakivi granites indicates that something has worn their corners off. If this were a sedimentary rock, physical abrasion might be invoked, but erosive tumbling is unlikely in an igneous rock formed from a melt. Instead, the rounding may have been caused by the injection of a second melt of different composition into the magma chamber as crystallization was occurring, causing the earlier formed orthoclase crystals to be partly resorbed into the magma, then overgrown by a rind of albite.

If rapakivi seems little more than an arcane entry in the annals of Finnish geology, consider that this sort of "magma mixing" is now thought to be implicated in initiating explosive eruptions like those of Vesuvius, Krakatoa, Pinatubo, and the Yellowstone caldera. Granitic magmas, rich in silica (SiO_2), are extremely viscous and sluggish and, left to their own devices, might simply solidify in the subsurface. It may take a thermal kick by a hotter basaltic magma to get them moving, with cataclysmic results. Rapakivi granites, with their strange round crystal balls, have something to tell us about what happens in a magma chamber just before the mayhem.

See also Granitization; Grus; Nuée Ardente.

S cree
Slippery slopes

From an Old Norse word meaning "slide" or "glide," **scree** refers in aggregate to loose rocks that mantle the side of a cliff or mountain. Scree-blanketed slopes can be quite treacherous because the stones within them are often stacked at the angle of repose—the steepest inclination that a granular material can maintain. This angle, typically in the range of 30–40°, is surprisingly independent of the size of the pile or the constituent pieces: a small mound of sugar on a plate can support the same maximum slope as a heap of boulders on a mountainside (about 35°). The *shape* of individual rocks is a more important determinant of the angle of repose, because this affects the way the pieces interlock frictionally. Typically, sharp, broken fragments can sustain a higher slope than less angular ones.

Although "repose" suggests a state of calm stability, a scree slope at the angle of repose is in fact everywhere at the point of failure. Any perturbation to it can launch a cascade of adjustments that act to restore equilibrium. In some cases, a single footstep can trigger avalanches of various sizes both below and, most dangerously, *above* an unsuspecting hiker.

An upside to the intrinsic instability of scree, at least on slopes with inch-sized rock fragments, is that it makes possible the exhilarating sport of "boot-skiing," a secret thrill practiced by many generations of field geologists. A mountain that took hours to climb can be descended in long, gliding strides in a matter of minutes, while crying "screeeeee"!

See also Breccia; Grus; Pedogenesis.

Skarn

A slow roast

Although it sounds like something that might have been served at a Viking banquet, ***skarn*** is an old Swedish miners' term for a particular metamorphic mineral concoction. Skarns require certain specific ingredients—but given the idiosyncrasies of particular geologic "kitchens," they rarely turn out exactly the same way twice.

To make a classic skarn, you should use approximately equal parts of a clay-rich rock like shale and a carbonate like limestone or dolostone. You'll also need a heat source such as a granitic intrusion. Let the hot magma bake the sedimentary rocks for at least a few tens of thousands of years. Once they reach about 800°F, the sedimentary rocks will begin to let go of volatiles they acquired at Earth's surface and have stored in their

constituent minerals: clays will discharge water, carbonates exhale carbon dioxide. (Heating carbonate rocks for skarn releases some greenhouse gases, but the amount is small compared with that released annually from limestone roasted in the production of concrete.) The liberated fluid phases, in turn, will help to redistribute soluble elements around the rock mass, fostering the formation of a completely new array of extraordinary minerals from the original ingredients.

Once it's done, your skarn might be dotted with ores of copper, lead, molybdenum, tungsten, and tin, but sometimes, especially if the baking is incomplete, you'll just end up with a lot of talc. And take care, because you might cook up a nasty form of asbestos called "tremolite." Many other exotic minerals will appear at higher temperatures if you are patient.

The sequence in which skarn minerals form with increasing temperature was first worked out in experiments conducted in the 1930s by Norman Bowen at the Geophysical Laboratory of the Carnegie Institution in Washington, DC; he also did groundbreaking work on the origin of granite, disproving the misguided concept of granitization. As Bowen was baking up skarns in his lab in Washington, he was clearly aware of world events, and he composed an astonishing mnemonic device that was both a powerful political statement and a way to remember the sequence in which skarn minerals appear. Bowen turned the minerals tremolite, forsterite, diopside, periclase, wollastonite, monticellite, akermanite, spurrite, merwinite, and lawsonite into a short and chilling work of poetry: "Tremble, for dire peril walks, monstrous acrimony spurning mercy's laws."

Bowen was, of course, referring to Hitler and his army, but the warning could apply to ruthless marauders from any era in human history.

See also Acasta Gneiss; Amethyst; Eclogite; Granitization.

Slickensides
Science friction

Slickensides is the onomatopoeic term for streaks or striations on ancient fault surfaces. There is something vaguely comical about the word, which has been used by geologists since the late 18th century; it seems a natural candidate for a tongue twister.

Like the stock characters in old Westerns who tell the sheriff, "They went thataway," slickensides point in the direction of fault slip. They occur as several distinct "species" that reflect different processes of formation along a slipping fault. Some slickensides are scratches

or grooves on highly polished, even shiny, rock surfaces; these are the product of abrasive wear along a fault. Others, more precisely called "slickenfibers," are defined by sets of elongated crystals—often quartz or calcite—that lie parallel to each other on a fault surface. These form by the incremental precipitation of minerals from deep groundwater that was pumped through the fault zone with each episode of slip.

Studying slickensides in rocks that have been "exhumed" by erosion yields insights into the behavior of faults at depths that are otherwise inaccessible. Their orientations make it possible to reconstruct tectonic stress fields, and slickenfiber minerals bear information about the interplay between fluid flow and fault slip over the course of many earthquake cycles.

Grammatically, "slickensides" is a noun that is almost never used in the singular, since slickensides stick together in frictional factions on fractures.

See also Breccia; Mylonite; Pseudotachylyte.

Speleothem [SPEE-lee-oh-them (with unvoiced "th")]
Drip feed
From the Greek *spelaion* for "cave" and *thema*, meaning "something laid down"—like sediment or a treatise—a ***speleothem*** is any type of mineral deposit in a cave. Not all caves have them; speleothem formation requires specific oscillations in groundwater chemistry from acid to base and back.

Most caves are formed in limestone, a rock made mainly of the mineral calcite, which can be dissolved by weakly acidic groundwater. Calcite dissolution is most likely to occur close to the water table—the top of the

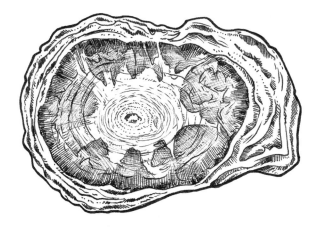

region in the subsurface where sediments and rocks are fully saturated—because groundwater there is at its most acidic, having assimilated carbon dioxide from the atmosphere and picked up more through interaction with organic material in the soil.

Acidic groundwater flowing through limestone quickly acquires a load of calcite (in the dissolved form of calcium and bicarbonate ions) and is thus buffered, losing its acerbic bite (it's no coincidence that over-the-counter heartburn tablets are made largely of calcite). When this mineral-laden water finds its way into the open space of a cave, where the air has much lower levels of carbon dioxide, the water will be out of equilibrium with its new surroundings. Like a high-altitude climber jettisoning items from her pack as the air gets thinner, the water gets rid of much of its cargo, exhaling carbon dioxide—and precipitating calcite on the cave walls as speleothem deposits.

Speleothems can have a wide range of forms, of which the most familiar are stalactites (rocky icicles hanging from a cave ceiling) and stalagmites (columns growing up from below a leaky spot), both from the Greek *stalagma*, "a drop." More complex morphologies, with a variety of evocative names, include curtains, ribbons, cascades, chandeliers, soda straws, moonmilk, popcorn, and cave bacon. Oddly shaped structures formed by geologic happenstance inspire fanciful descriptions and homespun lore that are inevitable features of commercial cave tours. I'm always impressed by the sheer ontological heterogeneity of these: an alligator might share a chamber with a bust of Napoleon; the coffin of a giant may lie, improbably, adjacent to the Liberty Bell.

The real stories that speleothems tell, however, are far more interesting. Protected from the erosive forces that prevail on the land surface, speleothems growing in caves quietly document events happening in the environment above and around them over time. In the central Italian Apennines, caves developed in a bone-white Jurassic limestone called the Calcare Massiccio (which, in Tuscany, where metamorphosed, forms Michelangelo's favorite marble) have markedly curved stalactites. These are essentially historical plumb bobs, recording the changing orientation of the vertical owing to the progressive tilting of the entire cave as the mountains keep rising and the crust continues to crumble.

The interiors of stalactites and stalagmites are even richer archives of information. Over time they grow, not only in length but in diameter, accreting concentric bands, reminiscent of tree rings, with the oldest material at the center. These can preserve a "memory," via their trace element and isotope geochemistry, of changes in climate and biological communities. Some speleothems constitute archives of environmental change going back more than a million years, making them valuable low-latitude counterparts to polar ice cores as sources of information about Holocene and Pleistocene climate variability. True to their name, speleothems are indeed "cave treatises"—dissertations, encyclopedias, whole libraries documenting forgotten worlds.

See also Firn; Karst; Varve.

Stygobite [STIG-oh-bite]
What lives beneath

Named for the mythical river Styx that carries souls to the underworld, ***stygobites*** are organisms that live only in subterranean groundwater environments—caves, deep rock fractures, and aquifers. Cut off from sunlight, subsurface organisms have nonetheless found ways to thrive.

One advantage of life underground is the relative constancy of conditions—no day and night, and no dramatic seasonal variations; even climate change may pass unnoticed. But just as geographic variations in soil type and climate give rise to diverse surface ecosystems, differences in underground habitats—especially the size of livable space and availability of water—create a surprising range of stygobite communities. There are microbes that eke out a living directly on underground rock surfaces, and tiny mites that sustain themselves on trace amounts of organic matter in water-saturated strata. Cave-dwelling fish, amphibians, crustaceans, and insects enjoy the roomy capaciousness of karst architecture. After accidentally stumbling or swimming into caverns and sinkholes, the ancestors of these organisms created shadowy versions of the ecosystems they had known at the surface deep in the Stygian darkness. Caves were probably the first shelters for our own species, and stygobites would have been bedfellows of our troglodyte ancestors.

While we're visiting the murky nether regions, we might make note of other underworld myths embedded in the geologic lexicon. These include plutons, igneous rocks crystallized underground and named for Pluto; allochthons, tectonically detached slabs of rock that allude to Pluto's "chthonic" realm; and the Hadean eon, Earth's first 500 million years, for which there is no native rock record, a Hades-like, hellish time too hostile even for stygobites.

See also Allochthon; Karst.

Stylolite [STY-lo-lite]
Marble marvels

A ubiquitous but often overlooked feature of carbonate rocks—limestone, dolostone, and their metamorphic equivalent, marble—***stylolites*** are surfaces along which partial dissolution has left dark, irregular selvages of insoluble trace elements like iron and manganese in the rock. Stylolites are what make marbles "marbled." But to understand their origin, we must visit the setting of their formation, preferably in scuba gear.

Most carbonate sediments accumulate in tropical waters from a continuous fine rain of biogenic material—
microscopic shells or "tests"—onto the seafloor, at water
depths unaffected by wave action. This process is in fact
Earth's long-term carbon sequestration strategy, the
main mechanism by which carbon dioxide exhaled by
volcanoes is removed from the atmosphere and stored
for geologic time spans, thereby preventing Earth from
becoming a runaway greenhouse planet (thank you,
carbonate-secreting organisms!). But that is not our
primary focus here.

Over time, as biogenic lime mud rains down to the sea-bed and is buried by more of the same material, it compacts, dewaters, and gradually becomes rock. As long as the accumulation process is uninterrupted, there should not be any visible layering in the resulting deposit—and yet most limestones and dolostones do look stratified. Their apparent bedding planes, however, are not actual depositional features but rather stylolites, "pseudobedding," formed as a result of compaction and dissolution of the carbonate minerals along horizontal surfaces.

Many minerals, and particularly calcite, are more soluble when under stress—a response that perhaps humans can empathize with. As groundwater moves through sediments, it will thus preferentially dissolve minerals along surfaces where they experience the highest stress. In the case of a deepening pile of carbonate mud, the greatest stress is from the vertical force of gravity, and so dissolution occurs along horizontal planes. Stylolites are not perfectly planar, however; for reasons that are still not fully understood, they have rough surfaces that look like zigzag stitching when seen in cross section.

Even if you've never noticed stylolites in limestones or dolostones, you've almost certainly seen them in marble—they create the wispy gray streaks that are mimicked even in bad Formica renditions. Marbles form when beds of limestone or dolostone find themselves caught up in mountain building and become recrystallized and distorted. The tiny carbonate crystals in the original rock reconfigure themselves into much larger ones, and this coarsening gives marble its characteristic milky translucency. (The same process, called "Ostwald ripening," occurs in ice cream that has been in the freezer too long.) Meanwhile, deformation exaggerates the amplitude of the jagged irregularities on the stylolite surfaces, creating the iconic look of marble.

I sometimes wonder whether Michelangelo, who spent hours with his nose close to stylolites in the luminescent Carrara marble, ever pondered their origin. Like many phenomena in nature, stylolites are so ubiquitous we hardly give them a thought. Yet in their absence, our lives would be subtly but significantly impoverished.

See also Karst; Speleothem.

Sverdrup [SVARE-drup]

Current affairs

Named for the eminent Norwegian oceanographer Harald Ulrik Sverdrup (1888–1957), a *sverdrup* is the unit used to quantify the volumetric rate at which an ocean current moves seawater, defined as one million cubic meters per second. The sverdrup lies at the opposite end of the hydrological spectrum from the darcy, a unit used to describe groundwater flow. The best way to grasp the enormity of this measure is to imagine a vertical "doorway" in the ocean whose height and width are both one kilometer—more than half a mile—and through which water is flowing at one meter per second (2.2 miles per hour). That is a sverdrup.

Where the Gulf Stream enters the Atlantic off the southern tip of Florida, it transports water at the rate of about 30 sverdrups. As it moves northward, it steadily picks up speed, and by the time it nears Newfoundland, it is surging at 150 sverdrups. To put this in perspective, even in the most extreme flood years, the Mississippi River—which drains 40 percent of the land area of the continental United States—delivers water to the Gulf of Mexico at a rate of only 87,000 cubic meters per second or 0.087 sverdrups. The Gulf Stream varies seasonally by more than 5 sverdrups.

The Gulf Stream and other currents convey not only immense volumes of water but also a vast amount of heat, in a ceaseless effort to equalize temperature differences between the tropics and the poles. As a result, the oceans are critical modulators of Earth's climate system. An added complexity is all the salt that ocean currents carry, which together with temperature determines the density of seawater. On their journey north, the waters of the Gulf Stream become progressively cooler, and also, as a consequence of evaporation, saltier. Both of these factors increase the water's density. When it reaches the latitude of Iceland, this incoming water is denser than the local ambient water, and it sinks to the seafloor, returning as a deep bottom-flowing current, a salty leviathan, to lower latitudes. This convective overturn of the ocean is called the *thermohaline* (heat-salt) circulation system, and disruptions of it have been responsible for abrupt climate change in the recent geologic past.

To block a torrent as mighty as the Gulf Stream requires an equally powerful aqueous force—namely, volumes of freshwater delivered at rates measured in sverdrups. Although ordinary rivers like the Mississippi aren't capable of that, the collapse of an ice sheet or catastrophic drainage of a meltwater lake (see *Jökulhlaup*) could be. Several large-magnitude events of this kind are thought to have occurred at the very end of the Pleistocene around 12,000 years ago.

At this time, Earth was emerging slowly from the long ice age; glaciers were relinquishing their grip, and vegetation was returning to mid-latitude lands. Then the warming world plunged precipitously back into glacial conditions for at least a millennium, an interval known as the "Younger Dryas" (named for a species of arctic flower that reappears in European peat from that time). The sudden chill seems to have occurred when the Gulf Stream became stalled as the result of dilution by huge volumes of fresh water from rapidly melting glaciers, thus starving the North Atlantic of heat. Climate records from Greenland ice cores show that the Younger Dryas ended as abruptly as it had begun; when the Gulf Stream finally reestablished itself after a thousand years, annual average temperatures in the Scandinavian region jumped by more than 10°F in a matter of decades. The Holocene had properly begun.

Climate scientists are concerned that accelerating rates of calving and melting from the Greenland ice cap due to human-accelerated global warming could again disrupt the Gulf Stream. At first this could, paradoxically, push Northern Europe into such cold conditions that agriculture would be impossible. In the early

2000s, this scenario was considered likely enough that the Norwegian government developed plans for how the country might sustain itself. More recently, however, climate models suggest that global temperature increases in coming decades may be so great that they would offset any cooling from the shutdown of the Gulf Stream. That's not exactly good news—especially since once the thermohaline system gets back into working order, there will be a whiplash to significantly higher temperatures.

Harald Sverdrup, the eponymous oceanographer, grew up in a village on the west coast of Norway where he developed a lifelong respect for the formidable energy of the sea. He spent his career studying and quantifying the physical forces that influence the powerful global system of ocean currents. He would likely have found it difficult to imagine that in the early 21st century, humans could be among them.

See also Darcy; Jökulhlaup.

Taphonomy [taff-ON-no-mee]
What becomes a fossil most

If history is written by the victors, the annals of prehistory were written by the vertebrates—as well as organisms with shells, exoskeletons, and other hard bits. Soft-bodied creatures may get written into the fossil record under certain unusual conditions that protect them from the depredations of oxygen—leading to the formation of special fossil beds called Lagerstätte, which provide rare glimpses of ancient ecosystems in their entirety. But most fossil-bearing strata contain only the tough, mineralized parts of organisms and therefore do not represent anything like a comprehensive census

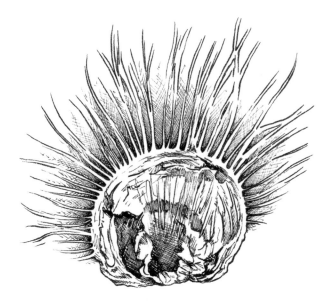

of the biosphere at any given time. Understanding and correcting for this intrinsic skewing of the fossil record is one aspect of ***taphonomy***, a subdiscipline within paleontology that focuses on the processes that lead to fossilization. The term comes from a Greek root meaning "grave" or "tomb."

It's not just particular organisms but whole environments that are underrepresented in the rock record. One of the most significant biases arises from the fact that the vast majority of sedimentary rocks accumulated in marine rather than land-based settings, for the simple reason that erosion, rather than deposition, prevails on land. Moreover, marine sediments accumulate with fewer interruptions than those in land environments.

As a result, patterns of evolution are far better documented for biota in the oceans than for ecosystems above sea level, and marine strata generally provide higher-resolution accounts of the great mass extinctions in the geologic past than do contemporary sediments deposited on land. For example, although the land-lubber dinosaurs are the celebrity victims of the end-Cretaceous extinction and get most of the media attention, the event is chronicled with far more nuance by the shells of tiny marine organisms called "foraminifera," who serve as the most reliable witnesses of the global cataclysm. The difference in coverage is a bit like the contrast between stories in the *National Enquirer* versus the *New York Times*.

Another, subtler taphonomic bias that has been documented only recently is the tendency for preburial degradation to simplify the apparent anatomy of organisms. This leads to a systematic tendency to assign fossils to older evolutionary "stem" groups and, more generally, to underestimate the biodiversity of ancient ecosystems.

Long-term changes in the chemistry of Earth's ocean and atmosphere have also affected the modes—and odds—of fossilization. Precambrian seawater had much higher concentrations of silica than do today's oceans. Chert, which is formed by precipitation of silica from water, is an ideal medium for high-fidelity fossilization, and the earliest microbial records of life are found in ancient cherts in Australia, South Africa, and elsewhere. In the modern oceans, siliceous phytoplankton called "diatoms," together with certain types of sponges, eagerly snap up silica from seawater, making bedded chert deposits rather rare. Instead, for the past few hundred million years, limestone—itself a largely biogenic rock produced by calcifying organisms—has eclipsed chert as the main sedimentary host for fossil preservation. So the rules about how to become a fossil have themselves evolved with Earth and life.

While the distortions inherent in the fossil record may seem insurmountable obstacles to understanding life of the past, recognizing that the record is incomplete has inspired paleontologists to think creatively about how to fill in the gaps. Knowing how energy flows through the trophic levels of modern food chains, for example, makes it possible to infer something about who was left out of the geologic chronicles. And even the organisms that didn't themselves become fossils sometimes left clues about their routines in the form of "trace fossils," such as crawling tracks and burrows.

Adopting a taphonomic frame of mind might in fact serve us well in thinking about history. By accepting the fact that the official accounts have been censored, we can begin to imagine who the victors have written out, and why. And we can aspire to keeping more complete and representative records for future generations.

See also Bioturbation; Lazarus Taxa; Tully Monster.

Thalweg [TALL-vegg]

The valley below

Recollecting his early days as a Mississippi riverboat pilot, Mark Twain wrote, "I went to work now to learn the shape of the river; and of all the eluding and ungraspable objects that ever I tried to get mind or hands on, that was the chief." But Twain did in fact grasp something essential about the shape of the Mississippi, and other rivers: that they defy simple geometric description—not only in map view but also in three dimensions, not to mention over time.

Rivers just can't resist meandering; they zig and zag in increasingly exaggerated turns, nearly looping back on themselves, like old-fashioned ribbon candy, until they finally lop off a bend and leave an oxbow lake behind. Meandering is such a deep physical instinct that even streams confined to straight, engineered channels start snaking once they establish beds of sediment.

As the water's momentum shifts from bank to bank, the outer arc of a river bend is eroded, creating a "cut bank," while the quieter inner arc is the site of deposition, forming a "point bar." Any good riverboat captain knows that the depth of the river is greatest next to cut banks, not in the middle of the channel. To avoid running aground, one must keep the boat on the path of the ***thalweg***, the winding line connecting points of greatest depth along the length of the river.

Thalweg is a 19th-century German neologism meaning "valley way" and was in widespread use in the English scientific lexicon by 1900. Although there is no evidence that Twain ever used the term, his pen name is a direct allusion to finding it: "mark twain" was a reassuring call by a steamboat leadsman that the water was two fathoms deep and safe for passage.

A river's thalweg can be of political, as well as navigational, importance. Where a river marks the boundary between provinces, states, or countries, there is a legal principle, the "Thalweg Doctrine," that designates the thalweg as the official position of the border. Recent disputes between Iran and Iraq and between India and

Bangladesh have been resolved, provisionally, by invoking this principle. But no river is ever content with its shape. They all endlessly reconfigure themselves, elusive and ungraspable, confounding our desire for permanent geographies and predictable geometries.

See also Jökulhlaup; Pedogenesis.

Thixotropy [THICK-suh-trope-ee]
Call me quick

Once an essential source of dramatic tension in adventure movies, quicksand is now like a matinee idol whose fame has dimmed. Given its association with cringeworthy colonialist plots and stereotyped characters, one might imagine that quicksand was just another expression of unenlightened ideas about "exotic" parts of the world. But in fact, quicksand is real, and the phenomenon that underlies it, *thixotropy*, or sediment liquefaction, is subtle and fascinating.

In Hollywood quicksand scenes, the more that a hapless victim struggles, the deeper he or she sinks. A villain may be swallowed up entirely, while a virtuous character is typically saved by a cooler-headed hero who exhorts the flailing party to be still. The social and moral implications of this may be problematic, but the physics of these scenarios is basically correct: A water-saturated sediment that is firm in its resting state, with its constituent particles in contact with each other, can change in an instant to a liquid slurry when jarred by a footfall. Even a small shock to a thixotropic mixture can cause the load-supporting scaffolding of particles to collapse, and the material loses its strength. Concrete is a familiar example of a thixotropic medium; cement mixers must

continuously churn to keep it in a liquefied state until it is ready to be laid down.

In nature, thixotropy happens more often in fine-grained clays than in sand. "Clay," which is both a grain size and a mineral group, is a complex material in which the very large net surface area of the tiny particles and the molecular structure of the minerals conspire to create unusual bulk properties. Flat clay particles, with un-bonded electrons all around their edges, are attracted to each other electrostatically and this creates microscale structures that give it macroscopic strength. But these miniature houses of cards are easily demolished by physical disturbance.

In addition, many clay minerals—the smectite group or "swelling" clays—have open crystal structures that allow them to absorb large amounts of water. Such clays tend to have very large differences between their resting and disturbed strengths and are described as "sensitive" or "quick." Their abrupt loss of cohesion is the underlying cause of many devastating landslides. But their capacity to restore their strength can also be dangerous: After stepping into a quick clay that rapidly becomes firm again, a person may find it impossible to pull his feet out. If this happens on a muddy tidal flat, the situation can be deadly, but not for the reasons depicted in the old movies.

Quicksand pockets in the middle of deserts, a common cinematic trope, are very rare, for the obvious reason that water is central to the whole phenomenon of thixotropy. One circumstance in which large bodies of sand may undergo liquefaction is in an earthquake. When shaken by seismic waves, the grains in a water-saturated sand deposit can go into suspension and gush out onto the surface in what are called "sand boils." Over time, such sand eruptions may be buried by organic-rich soil matter that can be carbon-dated and can thus provide information about the frequency of large earthquakes in tectonically active areas and improve seismic risk assessments. In this way, quicksand, the onetime menace of the silver screen, may actually help save lives.

See also Lahar; Lusi; Pedogenesis; Turbidite.

Tiktaalik [tick-TAH-lick]
Fish out of water

For years I've kept a folder of geologically themed cartoons clipped from newspapers and magazines that I like to sprinkle into classroom lectures to add a bit of levity. Dinosaurs and woolly mammoths are recurrently featured, but by far the most common theme in these single-panel sketches is the moment in geologic time when an adventurous fish ventured onto land.

In one panel, a mother fish, still in the water, yells to her prodigal son crawling onto shore: "Come back here this instant, young man!" Another depicts a pair of fish clambering up a sandy beach, and the trailing one asks, "Are you sure we don't need a visa or something?" And in one of my favorites, a particularly exuberant proto-tetrapod exclaims, "Walk, hell—I gotta dance!"

In reality, of course, the emergence of vertebrates onto land didn't happen on a single sunny afternoon but was instead a process that unfolded gradually—in specific, during the late Devonian period between about 390

and 375 million years ago. The Devonian was the Age of Fishes, with more fish diversity in the world's oceans, as measured by the number of taxonomic classes, than exists today. Some members of one group of lobe-finned fish, the sarcopterygians (Greek for "flesh-winged") took to living in shallow inlets and estuaries, sometimes far inland from the open ocean. And one fine geological day, one of these did leave the water and founded the lineage that led to amphibians, reptiles, and mammals—including us.

Paleobiologists believe that the impetus to exit the marine realm was not adolescent recalcitrance or an irresistible urge to dance but a period of widespread ocean anoxia. Even with gills, it may have become easier to extract oxygen from the air than from oxygen-poor water. Although it might seem that the great evolutionary leap (so to speak) of this time would have been to turn fins into feet, sarcopterygians were likely using their fins to "walk" along the muddy bottom long before any of them became amphibious.

In 2004, on remote Ellesmere Island in High Arctic Canada, a group of paleontologists discovered the key transitional fossil, the now-famous ***Tiktaalik*** (Inuktitut for "shallow water fish"). *Tiktaalik*, found in strata dated at 375 million years old, has such a perfect blend of fish and tetrapod characteristics that Neil Shubin, coleader of the expedition, describes it as a "fishapod." And *Tiktaalik* revealed that the real anatomical innovation that turned fish into pedestrians weren't trotters or toes but the development of a neck that separated the head from the rest of the body. Fish don't need necks, but for land-based creatures, a neck is critical for mobility and range of vision. *Tiktaalik* literally stuck its neck out and changed the world.

The anatomical novelty introduced by *Tiktaalik* also opens a world of new possibilities for cartoonists: A throng of *Tiktaalik* is scrambling up the beachfront and, for the first time in Earth's history, one says, "Stop breathing down my neck!"

See also Nunatak; Tully Monster; Simplified Geologic Timescale (Appendix 1).

Tully Monster
Alien autopsy

Even if you're not a *Star Wars* groupie, you're probably familiar with the iconic "cantina scene" set in a rowdy tavern in the spaceport town of Mos Eisley. It's clearly a rough place, and yet there is also an underlying ethos of tolerance—the bar is crowded with anatomically diverse citizens of various planets who seem unfazed by each other's bulging eyes, lurid green skin, or pendulous proboscises. I think of that scene—and the jazzy alien musical score—whenever the paleontological mystery of the ***Tully Monster*** comes up in conversation (as it does now and again among geologists).

In 1955, an amateur fossil collector named Francis Tully was nosing around for specimens in the famous Mazon Creek fossil beds in northeastern Illinois, an Upper Carboniferous (310 million year old) Lagerstätte rock unit that contains exquisite fossils of plants and rarely preserved soft-bodied animals. Many of the Mazon Creek fossils occur as concretions—ellipsoidal nodules that weather out of the rocks. Tully cracked one of these open and found himself face-to-face with a

bizarre creature that appeared to have been assembled from random spare parts. Its body was like a sausage that flattened slightly into a rudimentary tail. Although it lacked a distinct head, it had two eye stalks that stuck out like toothpicks twice the width of its body. And at the front, a hose-like appendage equally as long as the sausagey back-end terminated in what might have been a mouth but resembled a toothed castanet. Even though the creature was only about four inches long, it was monstrous.

Tully brought the specimen to the illustrious Field Museum in Chicago, but no one on the paleontological staff there could identify it or even say for certain whether it was an invertebrate or vertebrate animal. More than

60 years later, the controversy over the proper taxonomic classification of the Tully Monster rages on. The enigma seemed to be resolved in 2016 when a group of British researchers published a paper in the journal *Nature*, making the case that the Tully Monster was a primitive vertebrate, most closely related to hagfish and lampreys. But other groups continue to argue that Tully was a spineless thing, perhaps an arthropod or a tunicate (a group that includes modern sea squirts).

So for now, although the Tully Monster has a scientific genus and species name, *Tullimonstrum gregarium* (the latter part indicating that a "herd" of other specimens have been found at Mazon Creek), it is officially assigned only to the kingdom Animalia and within that, the "clade" or subgroup Bilateria, which includes all animals that have body plans with mirror symmetry.

Or maybe Tully, with its long schnoz for trumpeting and its clacking mouth on percussion, was a member of a Mos Eisley band that got stranded on Earth after a late Carboniferous gig.

See also Ediacara; Taphonomy; *Tiktaalik*; Simplified Geologic Timescale (Appendix 1).

Turbidite [TER-bid-ite]
Muddy waters
A rock with a turbulent past, a ***turbidite*** is the deposit left by a fast-moving slurry of continent-derived sediment flowing out onto the deep ocean floor. Like ophiolites, turbidites were mapped and described more than a century before their origin and significance were understood. And their role in Earth's long-term tectonic evolution has only recently been fully appreciated.

Turbidites are typically found folded or tilted on edge in mountain belts as thick repetitive sequences of light-colored sandstone layers overlain by darker siltstones and shales. Alpine geologists in the mid-19th century called these deposits "flysch," a Swiss German term that may come either from the verb *fliessen* (to flow) or from *Fleisch* (meat or flesh): to hungry field geologists, the stripy look of these was perhaps reminiscent of bacon. Contemporary English geologists, meanwhile, called such strata "greywacke," an adaptation of the German term "grauwacke," meaning simply "gray stone."

By the late 19th century, flysch or greywacke sequences had been mapped in the Appalachians, British Caledonides, and Italian Apennines (and were integral to the long-abandoned geosyncline theory for mountain building), but geologists could point to no modern process that would produce such great accumulations of rhythmically bedded strata. This was a troubling violation of the doctrine of uniformitarianism, the foundational principle for the young science of geology: the precept that all rocks and geologic features formed in the past should be explicable by reference to phenomena observable in nature at present.

The enigmatic process of flysch formation finally revealed itself one day in November 1929, when a magnitude 7.2 earthquake occurred off the south coast of Newfoundland, far beneath the waters of the then prolific Grand Banks cod fishery. Less than a minute after the earthquake, 12 telegraph cables were cut in rapid succession by something that had swept across the seafloor at up to 60 miles per hour. Two hours later, a series of tsunami waves devastated fishing villages in the

narrow inlets of Newfoundland's Burin Peninsula, killing 28 people. It was later recognized that the earthquake had triggered an immense submarine landslide, which, in turn, not only generated the tsunami but also unleashed the mysterious entity that had cut the telegraph cables—a roiling juggernaut of sediment careening off the continental shelf onto the deep seafloor: a turbidity current.

The complex physics of turbidity currents was soon being explored by geologists in scale-model flume experiments. These showed that the characteristic, repetitive sandstone-siltstone-shale motif in flysch sequences is the signature of a turbidity current progressively jettisoning its sediment load as it loses momentum, starting with the heaviest particles. And this meant that the immense thicknesses of flysch—now renamed "turbidites"—in mountain belts around the world were records of countless thousands of events, such as the Grand Banks earthquake and landslide in 1929. However, the fact that these deep-sea sediments occur high in montane regions was not satisfactorily explained until the emergence of plate-tectonic theory in the 1960s finally provided a mechanism for crustal deformation and uplift.

At least one group of organisms, a genus of bivalve mollusks, is dependent on turbidity currents for survival. The xylophagaidae, or "wood eaters," subsist entirely on fragments of wood that make the improbable journey from land environments to the deep sea via turbidity currents.

Most recently, geologists have begun to recognize the surprising role turbidity currents play in keeping Earth on an even keel over geologic timescales: specifically, they are the critical step in recycling continental crust. Earth's oceanic crust is disposed of quite easily: born at mid-ocean ridges by partial melting of the mantle, it returns to the mantle via subduction about 150 million years later when it is older, colder, and denser. But continental crust, even when it is very old, is too buoyant to be subducted into the mantle directly. And erosion can't completely get rid of continental crust, because most eroded sediments end up being deposited on the continental shelves, which, although they are underwater, are still underlain by continental crust.

Turbidity currents, however, are the one means by which continental material can end up on the deep ocean floor where it may be subducted along with the underlying oceanic rocks. In this way, turbidity currents "close the loop" for continental crust. Fortunately, some turbidites have escaped this fate and have been pushed back to their ancestral home on the continents by the tectonic forces that build mountains, making it possible for geologists to ponder them.

See also Benioff-Wadati Zone; Geosyncline; Lahar; Uniformitarianism.

Twist Hackle

Give me a break

If you've ever picked up shards from a broken window or dish, you've probably seen the texture called *twist hackle*, a common feature on fracture surfaces. It's also likely that you never noticed it; after such mishaps, few people pause to consider precisely how an object was reduced to smithereens, but if one cared to, it would be possible to recreate the process of fragmentation in detail.

An entire subfield within material science, fractography, is concerned with decoding cracks by reading the traces they leave as they rip through a brittle medium. Fractographic analysis is important in forensic cases involving the failure of metallic components (e.g., airplane fuselages). Although a catastrophic breakdown might take only a split second, the chronology of fracture formation and propagation can be reconstructed (and culpability sometimes determined).

Geologists have adopted the methods of fractography for use in interpreting ruptures in rocks, not to assign blame but to better understand the limits of rock strength—essential for building safe roadcuts, bridges, dams, and other types of infrastructure.

To read a rock fracture, one must first determine which of two broad types it is, since each has its own characteristic idiom. The two types are distinguished by the geometric relationship between the fracture surface and the direction of motion. Shear cracks involve displacement parallel to the fracture surface. Once these become large enough, they are called "faults," and depending on whether they fail seismically or by slow incremental creep, they may develop a variety of slip-related features, including breccias, slickensides, mylonites, and pseudotachylytes.

The other type of fracture is a tensile crack, which gaps open in the direction perpendicular to the fracture surface. This is the main mode of failure for brittle objects like windows and crockery, as well as for rocks in the shallow subsurface. Once formed, a small tensile crack tends to grow in length, like a "run" in hosiery, because of the high concentrations of stress at its tips. This is why engineers are careful to load brittle material in compression whenever possible, so that any nascent tensile cracks are pressed closed.

Some tensile fractures propagate at a leisurely pace— perhaps only a fraction of an inch per day. A familiar example is a small crack in a windshield that starts at a divot from a flying pebble and then makes its way across the entire window over days or weeks. Such a crack is called "subcritical" because it propagates more slowly than the speed of sound in the medium. The surfaces of these cracks are generally smooth and featureless.

"Critical" cracks, in contrast, hurtle through the material at acoustic speeds, and they leave behind telltale signs of the violence of their passage. A critical crack starts as a flat, lentil-shaped opening that rapidly expands outward, gaining speed as it grows, and the texture of the fracture surface becomes correspondingly rougher with distance from the origin. Fractographers have developed an evocative, even poetic, set of terms for these concentric zones. The smooth "mirror" gives way to the slightly rougher "mist," then to rippled and branching "plumose structure." Finally, as the crack encounters some other preexisting surface, such as a bedding plane, it fans out into a series of "en echelon" segments, forming the ragged fringe called "twist hackle."

Happily, one can practice fractography without the trauma of a smashed window or shattered heirloom. Miniature versions of twist hackle can be produced quite nicely by snapping a piece of baker's chocolate or hard cheese in half. And after the experiment is finished, everything can be cleaned up safely in a couple of bites.

See also Breccia; Mylonite; Pseudotachylyte; Slickensides.

Unconformity

Conspicuous absence

In 1923, when a reporter asked British explorer George Mallory why he'd want to scale Mount Everest, Mallory famously, and flippantly, replied: "Because it is there." The *thereness* of mountains would seem to be beyond dispute. But to geologists, mountains are ephemeral features that grow, exist for a time, and then are erased by erosion. In fact, the beginnings of modern geology can be traced to observations by a British explorer long before Mallory's time about mighty peaks that *weren't* there. Call it the "Case of the Missing Mountains." Its resolution led to the discovery of Deep Time.

Around 1780, Scotsman James Hutton, a physician, gentleman farmer, and eccentric polymath, had formulated a remarkably modern theory of how the solid earth worked, based as much on his fervid imagination as sober

observation. He was convinced, for both theological and geological reasons, that the destructive force of erosion must be counterbalanced by processes of topographic rejuvenation. Hutton was among the first natural scientists to recognize evidence for intrusive magmas (see *Granitization*), and he correctly surmised that the same subterranean heat source that melted rocks might also power crustal deformation and mountain building.

So Hutton's mind was primed to understand the significance of a now celebrated outcrop called Siccar Point on the east coast of Scotland, near the border with England, which he happened upon while boating with friends in 1788. Two distinct sequences of sedimentary rocks are found there: a lower sequence with beds standing at a near vertical orientation and an upper one with beds gently inclined. It was the rubbly, irregular surface between these two sequences—now called an **unconformity**—that caught Hutton's attention.

In a remarkable leap of inductive thinking, Hutton realized that the steeply inclined rocks were the remains of a mountain belt, in which the strata had been upended and crumpled. He also understood that the surface between the tilted layers and the overlying rocks represented the time it would have taken for those mountains to be eroded away, far longer than the 6,000 years biblical scholars had suggested for the entire age of the earth. Without giving it a formal name, Hutton instinctively practiced uniformitarianism—he thought it quite natural to invoke observable present-day processes like erosion in interpreting the rock record. At Siccar Point, Hutton had found his evidence for a self-renewing Earth, and he opened up vast eons for future geologists to explore.

Unconformities, or gaps in the geologic record of time, are in fact the rule rather than the exception; no single place on Earth has a complete and uninterrupted record of geologic time. Even the Grand Canyon, that mile-deep embodiment of Deep Time, has huge omissions. There are two major unconformities in the rock sequence of the canyon, recording immensely long intervals when erosion prevailed over deposition. The first occurs within the old Proterozoic rocks at the base of the canyon and represents an intermission of about 450 million years, more than enough time for the demolition and burial of an ancient mountain belt. The other—known as the "Great Unconformity"—separates those old rocks from the familiar flat-lying strata of the upper part of the canyon, and constitutes another 500 million years of silence. In places, the agents of erosion that formed the Great Unconformity—including glaciers, in the ancient ice age called the Cryogenian—"ate" into the older unconformity, so in total the gap denotes nearly a billion years.

An unconformity is the geologic equivalent of pages missing from a particular copy of a book, an interruption of the narrative in a given place. Fortunately, however, rock sequences at different places—copies kept in other geologic "libraries"—are missing different pages. Since the time of Hutton, the geologic timescale has been pieced together laboriously by combining the fragmentary, unconformity-riddled records from sites around the world into a single comprehensive volume, using fossils and isotopic dates as global "page numbers."

As products of erosion, unconformities are defined by absence—they are protracted pauses when rocks go silent. But through that very silence, they provide glimpses of the ephemeral: the fleeting contours of ancient land surfaces. Unconformities allow us to witness the work of forgotten rivers, terrains known to vanished creatures, and the vestiges of unnamed mountains that once were *there*.

See also Anthropocene; Cryogenian; Uniformitarianism; Zircon; Simplified Geologic Timescale (Appendix 1).

Uniformitarianism

Same as it ever was

An elaborate word for a rather simple idea, ***uniformitarianism*** is the foundational principle of geology: that in attempting to explain rocks, landscapes, and other records of the past, geologists should look to processes occurring on Earth today. In sound-bite form, it's usually summed up as "the present is the key to the past."

Adopting the practice of uniformitarianism was an essential step in establishing geology as a credible scientific discipline. In the early 19th century, fossil discoveries, the search for mineral ores, and the mapping of coal-bearing strata led to the proliferation of idiosyncratic and often untestable ideas about geological phenomena, many invoking divine interventions (especially Noah's flood) or cataclysmic processes never observed in historic times. Charles Lyell, an Englishman trained as an attorney but more passionate about studying the laws of nature, became concerned that the young science of geology would not mature properly if it continued to practice such "philosophical promiscuity."

Lyell took it upon himself to check this wantonness in his monumental three-volume treatise *Principles of Geology*, published between 1830 and 1833, which pre-

scribed the doctrine of uniformitarianism as an antidote
to such flights of fancy. Lyell's key argument, developed
over more than 1,000 pages, was that geologic interpretations should be based only on directly observable
phenomena. Lyell wrote so compellingly, even evangel-
ically, about the doctrine of uniformitarianism that generations of geologists adopted it as a sacred orthodoxy.
Over time, what Lyell had intended as a methodological
rule for practitioners of earth science—a sort of Occam's
razor for rocks—gradually became a conviction about
Earth itself. Believing that Earth was no different in the
past than it is today was one reason geologists rejected
for decades the idea that continents could move. And
maintaining that Earth had never experienced geologic
events beyond those of the kind witnessed in human history blinded geologists to evidence of large-magnitude
floods, supervolcano eruptions, extreme climate excur-
sions, and meteorite impacts.

Like so many "isms" that begin with reasonable principles but become rigid orthodoxies, uniformitarianism
has been both essential to, and dangerous for, geology.

See also Cryogenian; Jökulhlaup; Unconformity.

Varve
Dear diary

A *varve* is the sedimentary record of one year,
typically deposited in a glacial lake or fjord setting. The
term comes from a Swedish word meaning "turn" or
"cycle" ("solstice" in Swedish is *solvarv*—the turning of
the sun). A single varve includes a sandy to silty summer
layer representing the input from streams, and a thinner, clayey winter layer formed by the slow rain of fine

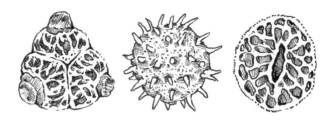

suspended particles when ice covers the surface of the water body. Like tree rings, varves provide an indirect but high-resolution archive of climate and environmental conditions over time.

In some Finnish lakes, there is a continuous annual varve record from the present back more than 10,000 years, to the start of the Holocene—the entirety of recorded human history. Within the varved sediments are pollen grains, fungal spores, insect parts, and trace elements that chronicle a landscape emerging from the Ice Age. The arrival of humans is easily recognized by an acceleration in sedimentation rates, a spike in lead levels, and a dramatic change in the mix of pollen types. Later entries in the varve archive include mercury from coal burning, radioactive isotopes from the Chernobyl disaster, and microplastics from household products. Varves have silently recorded it all, without comment or judgment.

See also Anthropocene; Cryogenian; Speleothem.

Xenolith
Wayfaring stranger
Fabricated by geologists from Greek components, the word **xenolith**, which shares a prefix with

xenophobia, literally means "foreign rock." More specifically, a xenolith is a stranger in unfamiliar igneous territory—a chunk of rock that happened to be picked up by an ascending body of magma and subsequently preserved within it as the magma crystallized. An analogy could be someone on a pier who got swept involuntarily onto a departing ship as passengers swarmed aboard—and then ended up residing permanently at the unintended destination.

Xenoliths have themselves been subject to a bit of xenophobia: British quarrymen who cut granite for building stones considered them unwelcome impurities and called them "heathens." Geologists, however, respect xenoliths. They are emissaries from otherwise inaccessible realms, carrying messages encoded within their minerals. In fact, most of the rare rock samples we have of Earth's mantle are xenoliths that were borne up either by basaltic magmas (as in Hawai'i, where such xenoliths create green sand beaches) or with the truly strange diamond source rocks called "kimberlites." These mantle xenoliths are the "ground truth" that make it possible to convert seismic wave data into inferences about Earth's interior.

In English words starting with "xeno," the "alien" connotation is salient, but in Greek, "xeno" can also mean "guest." This is how I prefer to think of xenoliths: as visitors who never really intended to leave home but, having come so far, are content to stay and share stories that will help us see the world differently. We should welcome them and listen.

See also Kimberlite; Komatiite; Mohorovičić Discontinuity.

Y

ardang [YAR-dahng]

Gone with the wind

Although it sounds like an oath that might be uttered by a pirate wrangling sails on the high seas, a *yardang* is a dry land—indeed, desert—phenomenon: an isolated, wind-sculpted rock outcrop surrounded by plains of sand. In a fortuitous alignment of the grammatical and the geological, "yardang" is the ablative form of a Turkish word meaning "steep bank." The ablative case, which doesn't exist in Germanic languages, is used to indicate motion away from an entity—particularly fitting for a landscape feature shaped by abrasive sandblasting.

Yardangs often have streamlined shapes that reflect prevailing wind directions: the side facing the wind is relatively steep and wide, while the lee side is gentler and tapering. Common in the dry and dusty equato-

rial region of Mars, yardangs shed light on long-term weather patterns there. Where the relentless Martian winds scour volcanic bedrock, they occur by the hundreds in vast, aligned arrays resembling great naval armadas—a sight that might prompt a pirate to exclaim "Yardang!"

See also Dreikanter; Erg; Haboob; Hoodoo.

Yazoo [YA-zoo]
Parallel lives

Named for a vanished Native American tribe and the river along whose banks they lived in what is now southern Mississippi, a *yazoo* is a small stream that flows near, and parallel to, a much larger one without joining it. This strange arrangement is rather like a quiet country road that runs adjacent to an interstate but has no entrance ramp for access to the major highway.

A yazoo stream develops when the natural levees of the bigger river—built over time by sediment settling out of floodwaters—become higher than the level of smaller adjacent streams. The yazoo, blocked from confluence with the river, must then make its own solitary way to the coast, even though the hustle and bustle of the main channel are so close by.

See also Pedogenesis; Thalweg.

Zircon [ZER-kun]
I will survive

How appropriate (even if it is an accident of alphabetical order) that the final entry in this volume is a mineral that can outlast virtually all other Earth materials.

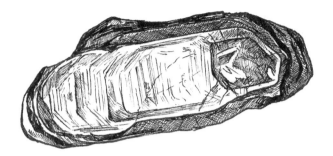

Zircon ($ZrSiO_4$), which occurs primarily as a minor constituent of granite, is singularly resistant to both chemical breakdown and physical abrasion. It forms at high temperatures, so it can endure metamorphic events that would recrystallize or even melt lesser minerals. A tiny crystal of zircon can be pummeled for millennia in pounding surf or tumbled for hundreds of miles along a riverbed—and come through unscathed. In contrast to diamonds—a form of carbon that is unstable at Earth's surface and slowly converts to graphite—zircons really are forever.

Zircon also happens to be ideally suited for isotopic dating: At the time of crystallization, it accepts some uranium (U) into the sites for zirconium (Zr) in its atomic lattice. Over time, the two isotopes of uranium, ^{235}U and ^{238}U, both radioactive, break down to their respective "daughter" isotopes of lead, ^{207}Pb and ^{206}Pb. Measuring the two lead-to-uranium ratios in a zircon crystal allows it to be dated with high precision. In other words, zircons not only have long memories but share them readily. The age of Earth's oldest-known rocks, the Acasta gneisses of Northwestern Canada, was revealed by grains of zircon. And the most ancient earth objects ever found are tiny zircon crystals that survive as grains in an ancient sandstone in the Jack Hills of Western Australia—but recall quite clearly their crystallization in a granitic magma 4.4 billion years ago.

In addition to being extraordinarily robust, zircon crystals also have the capacity for reincarnation. An old zircon crystal, after eons of dormancy, can grow new concentric layers when reheated in a magmatic or tectonic event. These resemble miniature tree rings—but may have formed in episodes of growth separated by hundreds of millions of years. A single zircon crystal can be a microcosm containing the tectonic history of a continent.

Perhaps the most remarkable zircon biography yet discovered is that of a crystal found in a granitic fragment in a moon rock collected in 1971 by Apollo 14 astronauts. The presence of any granite on the moon was surprising, and for years this was a puzzling anomaly. In 2019, zircon crystals from this fragment, dated at four billion years, were found to have trace element signatures that are completely different than those of any other lunar rocks—but very similar to Earth values. The astounding implication is that the chunk of granite, carrying its zircons, is in fact a rock that rocketed off the earth in a meteorite impact, was hurled into space, landed on the moon—and was picked up billions of years later by an astronaut who happened to be strolling by.

Similar Earth meteorites are probably strewn around the surface of Mars, and that thought gives me some strange comfort. In five billion years, when the lifetime of the sun has finally run its course, Mars may lie outside its Red Giant radius—but Earth will be engulfed. Perhaps on the bleak plains of Mars there will be a few surviving Earth-born zircon crystals that remember the halcyon days of this beautiful, bountiful, complicated, creative planet.

See also Acasta Gneiss; Amethyst; Cryogenian; Granitization; Kimberlite; Oklo; Simplified Geologic Timescale (Appendix 1).

Appendix 1

Simplified Geologic Timescale

Eon	Era	Period	Beginning (millions of years ago)	Geologic Highlights
Phanerozoic	Cenozoic	Quaternary	3	*Anthropocene* Holocene (last 10,000 yrs) Pleistocene (Ice Age)
		Neogene	23	
		Paleogene	65	Mammals diversify Giant birds
	Mesozoic	Cretaceous	140	*Dinosaur extinction* Atlantic Ocean opens
		Jurassic	200	First flowering plants Oldest surviving ocean crust
		Triassic	250	Age of the reptiles begins
	Paleozoic	Permian	290	*Greatest mass extinction* Pangaea formed *Panthalassa Ocean*
		Carboniferous	355	Widespread coal swamps *Tully Monster*
		Devonian	420	First amphibians *Tiktaalik*
		Silurian	440	Widespread coral reefs
		Ordovician	508	First land plants
		Cambrian	541	Modern animal phyla appear *Bioturbation ubiquitous*

Eon	Era	Corresponding time on Mars (see Areology)	Beginning (millions of years ago)	Geologic Highlights
Proterozoic	Neoproterozoic		800	*Ediacaran* organisms *Cryogenian* ("Snowball Earth") *Rodinia* supercontinent
	Mesoproterozoic		1600	
	Paleoproterozoic		2500	Oldest rocks in Grand Canyon *Oklo* reactor Great Oxygenation Event
Archean	Neoarchean	*Amazonian* *(continues to present)*	2800	Modern-style plate tectonics (subduction) Youngest *komatiites*
	Mesoarchean		3200	
	Paleoarchean	*Hesperian*	3600	
	Eoarchean	*Noachian*	4000	Oldest rocks on Earth: *Acasta Gneisses*
Hadean		*Pre-Noachian*	4500	No rocks from this period on Earth; known from *chondrite* meteorites, moon rocks, and a few Australian *zircon* crystals

Appendix 2

Entries by Language of Origin

Aboriginal Australian
Ediacara

Arabic
Erg
Haboob

Bantu
Oklo

Farsi
Namakier

Finnish
Rapakivi

French
Boudin
Nuée Ardente

German
Dreikanter
Firn
Klippe
Thalweg

Greek
Allochthon
Amethyst
Amygdule
Chondrite
Cryogenian
Geophagy
Katabatic
Panthalassa
Pedogenesis
Pneumonoultramicroscopic-
   silicovolcanoconiosis
Porphyry
Speleothem
Taphonomy
Thixotropy
Xenolith

Icelandic
Jökulhlaup

Inuktitut
Acasta
Nunatak
Pingo

Italian
Breccia

Javanese
Lahar
Lusi

Norwegian
Grus
Scree

Russian
Polynya

Slovenian
Karst

Swedish
Skarn
Varve

Tunica
Yazoo

Turkish
Yardang

Appendix 3

Entries by Topical Category

Fossils
Bioturbation
Ediacara
Gastrolith
Lazarus Taxa
Taphonomy
Tiktaalik
Tully Monster

Geophysics
Deborah Number
Eclogite
Geodynamo
Moho
Mylonite
Pseudotachylyte

Glaciers/Climate
Cryogenian
Firn
Jökulhlaup
Pingo
Polynya
Varve

Groundwater
Darcy
Karst
Speleothem
Stygobite

History of Geology
Benioff-Wadati Zone
Geosyncline
Granitization
Ophiolite
Turbidite
Unconformity
Uniformitarianism

Landforms
Erg
Hoodoo
Namakier
Nunatak
Pingo
Yardang
Yazoo

Metamorphism
Acasta Gneiss
Eclogite
Kimberlite
Skarn
Stylolite

Oceanography
Ophiolite
Sverdrup
Turbidite

Rivers and Floods
Jökulhlaup
Thalweg
Yazoo

Rocks and Minerals
Acasta Gneiss
Amethyst
Amygdule
Brimstone
Chondrite
Eclogite
Granitization
Grus
Kimberlite
Komatiite
Porphyry
Rapakivi
Xenolith
Zircon

Sedimentary Processes
Bioturbation
Erg
Lusi
Namakier

Stylolite
Thixotropy
Turbidite
Varve

Soil/Rock Weathering
Geophagy
Grus
Scree
Pedogenesis

Solar System
Areology
Chondrite
Nutation

Tectonics
Allochthon
Boudin
Breccia
Geosyncline
Klippe
Mylonite
Ophiolite
Panthalassa
Pseudotachylyte
Slickensides
Twist Hackle

Time
Acasta Gneiss
Anthropocene
Cryogenian
Oklo (natural nuclear reactor)
Panthalassa
Unconformity
Uniformitarianism
Zircon

Appendix 4

Entries by Geographic Location

Africa
GABON: Oklo
NAMIBIA: Ediacara
NORTH AFRICAN DESERTS: Erg,
Haboob
SOUTH AFRICA: Kimberlite,
Komatiite, Taphonomy

Antarctica
Firn

Arctic regions
Jökulhlaup, Nunatak, Pingo,
Polynya, Varve

Asia
HIMALAYA: Panthalassa
INDONESIA: Benioff-Wadati
Zone, Lusi, Rapakivi
JAPAN: Benioff-Wadati Zone
PHILIPPINES: Rapakivi

Australia
Acasta Gneiss, Ediacara,
Taphonomy, Zircon

Caribbean region
Nuée Ardente

Earth's interior
MANTLE: Benioff-Wadati
Zone, Eclogite, Kimberlite,
Komatiite, Mohorovičić
CORE: Chondrite, Geodynamo

**Extraterrestrial locations
(including fictional ones)**
MARS: Areology, Dreikanter,
Pedogenesis, Yardang
MOON: Acasta Gneiss,
Pedogenesis
MOS EISLEY: Tully Monster

Europe
ALPS: Allochthon, Ophiolite,
Panthalassa, Turbidite
BALKAN REGION: Karst,
Mohorovičić
ITALY: Brimstone, Nuée Ardente,
Rapakivi, Speleothem,
Stylolite, Turbidite

SCANDINAVIA: Ediacara,
    Rapakivi, Sverdrup, Varve
UNITED KINGDOM
    CORNWALL: Ophiolite
    ENGLAND: Jökulhlaup, Karst
    SCOTLAND: Allochthon,
        Granitization, Mylonite,
        Uniformitarianism

Mediterranean region
CYPRUS: Ophiolite
EGYPT: Porphyry
IRAQ: Namakier
TURKEY: Hoodoo

New Zealand
Geodynamo, Hoodoo,
    Ophiolite

North America
CANADA
    ALBERTA: Hoodoo
    CANADIAN ROCKIES:
        Allochthon
    NEWFOUNDLAND: Ediacara,
        Ophiolite, Turbidite
    NORTHWEST TERRITORIES:
        Acasta Gneiss
    NUNAVUT: Geodynamo,
        Nunatak, *Tiktaalik*
MEXICO
    YUCATÁN: Karst
UNITED STATES
    ADIRONDACKS: Boudin
    APPALACHIANS:
        Geosyncline, Turbidite
    ARKANSAS: Kimberlite
    ARIZONA: Haboob

BRYCE CANYON: Hoodoo
GRAND CANYON:
    Cryogenian,
    Unconformity
HAWAI'I: Amygdule,
    Komatiite, Mohorovičić
ILLINOIS: Tully Monster
LAKE SUPERIOR REGION:
    Amygdule
MAINE: Mylonite
MISSISSIPPI RIVER: Thalweg,
    Yazoo
MONTANA: Jökulhlaup,
    Klippe
NEBRASKA: Pneumonoultra-
    microscopicsilicovolcani-
    coniosis
WASHINGTON STATE:
    Jökulhlaup, Lahar
WISCONSIN: Mylonite
YELLOWSTONE: Rapakivi
ZION NATIONAL PARK: Erg

South America
COLUMBIA: Lahar

**Supercontinents of the
geologic past**
NUNA: Nunatak
PANGAEA: Panthalassa
RODINIA: Cryogenian

Appendix 5

Biographical References

**Benioff, Victor Hugo
(1899–1968)**
Benioff-Wadati Zone

**Bowen, Norman
(1887–1956)**
Granitization, Skarn

**Coney, Peter
(1929–99)**
Geosyncline

**Darcy, Henri Philibert Gaspard
(1803–58)**
Darcy

**Darwin, Charles
(1809–82)**
Geosyncline, Lazarus Taxa

**Hutton, James
(1726–97)**
Granitization, Unconformity,
    Uniformitarianism

**Kay, Marshall
(1904–75)**
Geosyncline

**Kelvin, Lord (William Thomson)
(1824–1907)**
Geosyncline, Granitization

**Lehman, Inge
(1888–1993)**
Geodynamo

**Lyell, Charles
(1979–1875)**
Uniformitarianism

**Mohorovičić, Andrija
(1857–1936)**
Mohorovičić

**Steinmann, Gustav
(1856–1929)**
Ophiolite

**Stille, Hans
(1876–1966)**
Geosyncline

**Sverdrup, Harald
(1888–1957)**
Sverdrup

**Tharp, Marie
(1920–2006)**
Eclogite

**Wadati, Kiyoo
(1902–95)**
Benioff-Wadati Zone

**Wegener, Alfred
(1880–1930)**
Benioff-Wadati Zone

**Werner, Abraham
(1749–1817)**
Granitization

Useful References

Alley, Richard. *The Two-Mile Time Machine: Ice Cores, Abrupt Climate Change and Our Future.* Updated edition. Princeton University Press, 2014.

Bjornerud, Marcia. *Reading the Rocks: The Autobiography of the Earth.* Basic Books, 2006.

———. *Timefulness: How Thinking Like a Geologist Can Help Save the World.* Princeton University Press, 2018.

Fortey, Richard. *Dry Storeroom No. 1: The Secret Life of the Natural History Museum.* Knopf, 2008.

———. *Earth: An Intimate History.* Vintage, 2005.

Garlick, Sarah. *Pocket Guide to Rocks and Minerals.* National Geographic, 2014.

Hazen, Robert M. *The Story of Earth: The First 4.5 Billion Years, from Stardust to Living Planet.* Penguin Books, 2012.

———. *Symphony in C: Carbon and the Evolution of (Almost) Everything.* W. W. Norton, 2019.

Knoll, Andrew. *Life on a Young Planet: The First Three Billion Years of Evolution on Earth.* Revised edition. Princeton University Press, 2015.

McPhee, John. *Annals of the Former World*. Farrar, Straus and Giroux, 2000.

National Audubon Society. *Field Guide to Rocks and Minerals*. Knopf, 1979.

Neuendorf, Klaus (author), and Jim Mehl (editor). *Glossary of Geology*. 5th edition. American Geosciences Institute, 2012.

Palmer, Douglas et al. *Earth: The Definitive Visual Guide*. 2nd edition. DK Smithsonian, 2013.

Shubin, Neil. *Your Inner Fish: A Journey into the 3.5 Billion-Year History of the Human Body*. Vintage Books, 2009.